Praise for

The Science of Yoga

"Sifts through the barrage of reports to distinguish many of the truths from the falsehoods. . . . A fluid writer, Broad brings his narrative to life with brisk portraits of people who combine yoga knowledge with science. . . . A well-researched book that belongs in the library of every yogi."

—*Yoga Journal*

"Popularly accessible, credibly researched, and breezily enjoyable. . . . Reading *The Science of Yoga* is like taking a carefree open-air automobile ride through yoga's fascinating past and its incredibly diverse present. . . . Broad guides the reader through the complex maze . . . sorting useful from faddish, dangerous from benign."

—*Kansas City Star*

"For anyone interested in yoga, [Broad] offers an objective, scientifically based study of the pros and cons of the ancient practice."

—*The Virginian-Pilot*

"An impressive chronology of yoga launches the book, from Indus Valley clay seals portraying figures in yogic postures, c. 2500 BCE, through current studies concluding that yoga lessens spinal deterioration and equals or surpasses exercise in reducing stress, improving balance and diminishing fatigue. . . . Broad's exploration of yoga's benefits, hype and hedonism lend a 21st century context to this most ancient of human pursuits."

—*Seattle Times*

"A much-needed analysis of a practice which is plagued by misinformation and blind faith."

—Urban Yoga

"Scrupulously documented, well written, interesting, and informative. Above all, it's fair."

—Greenville Yoga

"Yoga, an ancient practice with millions of modern practitioners, has been the subject of overheated speculation and grandiose claims; it has been dismissed without warrant as well, underappreciated by some who might well benefit from it. *The Science of Yoga* is a lucid and long overdue account of what scientists have found in their attempts to ferret out the truth about what yoga can and cannot do to heal and make better the body and mind. It is a fascinating and important book."

—Kay Redfield Jamison, Ph.D., author of
An Unquiet Mind and *Touched with Fire*

"*The Science of Yoga* offers a riveting, much-needed, and clear-eyed look at the yoga mystique. In this myth-busting investigation, science journalist William Broad applies his firepower as an investigative journalist to pull back the curtain on the little-discussed world of yoga injuries and risks, while setting the record straight about the numerous potential benefits. Downward Dog will never look the same."

—Daniel Goleman, author of *Emotional Intelligence*

"From a Pulitzer Prize—winning science writer, this fascinating book disentangles the truth about India's greatest contribution to self-care. Though he is hard on the commercialization of yoga, William Broad is optimistic and hopeful in pointing the way to its future as a major force in preventing and treating disease."

—Gail Sheehy, author of *Passages in Caregiving*

"Since its 'discovery' by the West in the nineteenth century, yoga has been constantly reinventing and repackaging itself, often as a panacea for all that ails Western man (and woman). In this compelling work of investigative journalism, William Broad exposes the 'scientific' claims made about yoga—from its much-vaunted healing powers to yogasms—to scientific scrutiny. *The Science of Yoga* is a wonderful read that any yoga practitioner thirsting for authenticity should study carefully before suiting up."

—David Gordon White, author of *Kiss of the Yogini* and
J. F. Rowny Professor of Comparative Religion,
University of California, Santa Barbara

"After reading *The Science of Yoga*, I am even more awed by the magnificent complexities of the human body and mind, and astonished that we can exert so much control over this invisible realm through the practice of yoga. Broad has not only thoroughly researched his topic, he has lived it."

—Alan Lightman, adjunct professor of Humanities at the
Massachusetts Institute of Technology and
author of *Einstein's Dreams*

Also by William J. Broad

*The Oracle: Ancient Delphi and the Science
Behind Its Lost Secrets*

Germs: Biological Weapons and America's Secret War
(with Judith Miller and Stephen Engelberg)

*The Universe Below: Discovering
the Secrets of the Deep Sea*

*Teller's War: The Top-Secret Story Behind
the Star Wars Deception*

*Star Warriors: A Penetrating Look into the Lives of the Young
Scientists Behind Our Space Age Weaponry*

*Betrayers of the Truth: Fraud and Deceit
in the Halls of Science*
(with Nicholas Wade)

THE SCIENCE

OF YOGA

The Risks and the Rewards

WILLIAM J. BROAD

Illustrations by Bobby Clennell

SIMON & SCHUSTER PAPERBACKS
New York London Toronto Sydney New Delhi

Simon & Schuster Paperbacks
A Division of Simon & Schuster, Inc.
1230 Avenue of the Americas
New York, NY 10020

First Simon & Schuster trade paperback edition December 2012

SIMON & SCHUSTER PAPERBACKS and colophon are registered trademarks of Simon & Schuster, Inc.

For information about special discounts for bulk purchases, please contact Simon & Schuster Special Sales at 1-866-506-1949 or business@simonandschuster.com.

The Simon & Schuster Speakers Bureau can bring authors to your live event. For more information or to book an event contact the Simon & Schuster Speakers Bureau at 1-866-248-3049 or visit our website at www.simonspeakers.com.

Designed by Renata Di Biase

Manufactured in the United States of America

10 9 8 7 6 5 4 3 2 1

The Library of Congress has cataloged the hardcover edition as follows:

Broad, William J.
  The science of yoga : the risks and the rewards / William J. Broad.
  —1st Simon & Schuster hardcover ed.
    p. cm.
  1. Hatha yoga. I. Title.
  RA781.7.B757 2012
  613.7'046—dc23   2011020408

ISBN 978-1-4516-4142-4
ISBN 978-1-4516-4143-1 (pbk)
ISBN 978-1-4516-4144-8 (ebook)

To Nancy
In Memoriam

There is no subject which is so much wrapped up in mystery and on which one can write whatever one likes without any risk of being proved wrong.

—*I. K. Taimni, Indian scholar and chemist, on the obscurity of yoga*

Contents

List of Illustrations

Main Characters

EZRA A. AMSTERDAM (1936–　). Cardiologist at the Medical School of the University of California at Davis. Led a 2001 study concluding that yoga improves aerobic conditioning.

BASU KUMAR BAGCHI (1895–1977). Scientist of Indian birth at the University of Michigan. Found that advanced yogis could slow but not stop their hearts.

KOVOOR T. BEHANAN (1902–1960). Yale psychologist born in India. Authored the 1937 book *Yoga: A Scientific Evaluation.*

HERBERT BENSON (1935–　). Cardiologist at the Medical School of Harvard University. Found that meditators reduced their breathing, heart rate, and oxygen consumption. Authored in 1975 *The Relaxation Response.*

T. K. BERA (1949–　). Director of research at Kaivalyadhama, the scientific ashram of Gune in the hills south of Bombay. Found advanced yogis skilled at slowing their metabolisms.

GLENN BLACK (1949–　). Yoga teacher and bodyworker. Instructed at Omega Institute in Rhinebeck, New York. Spoke openly of yoga injuries.

THÉRÈSE BROSSE (1902–1991). French cardiologist. Showed that advanced yogis could slow their heart rate and blood flow.

LORI A. BROTTO (1975–　). Sex researcher at University of British Columbia. Reported in 2002 and 2009 that fast breathing can result in sexual arousal.

MAYASANDRA S. CHAYA (1953–　). Indian physiologist. Led team reporting in 2006 that yoga lowers the resting metabolic rate of practitioners and does so twice as effectively in women as in men.

BIKRAM CHOUDHURY (1946–　). Yogi entrepreneur. Born in Calcutta and based in Los Angeles. Founded Bikram hot yoga. Set up hundreds of licensed and franchised studios around the globe.

MAIN CHARACTERS

CAROLYN C. CLAY (1980–). Sports scientist at Texas State University. Led 2005 study reporting that yoga has few cardiovascular benefits.

KENNETH H. COOPER (1931–). Physician who coined the term *aerobics* and advocated energetic sports. Reported few cardiovascular benefits from calisthenics, isometrics, and low-impact exercise.

JAMES C. CORBY (1945–). Physician at Stanford University School of Medicine. Led team reporting in 1978 that people in Tantric meditation undergo a number of physiologic arousals.

INDRA DEVI (1899–2002). Actress turned yogini. Studied with Gune and Krishnamacharya. Taught in Hollywood, Russia, and Argentina. Popularized yoga in her 1953 book *Forever Young, Forever Healthy*.

VIKAS DHIKAV (1974–). Medical doctor at Ram Manohar Lohia Hospital, New Delhi. Led team reporting in 2010 that men and women who take up yoga enjoy wide improvements in their sex lives.

CTIBOR DOSTÁLEK (1928–2011). Czech neurophysiologist. Studied advanced yogis and reported that their brains exhibited waves of excitement indistinguishable from those of lovers.

GEORG FEUERSTEIN (1947–). Indologist of German birth. Authored or coauthored more than thirty books, including *Yoga for Dummies*. Served as editor of *International Journal of Yoga Therapy*.

LOREN FISHMAN (1940–). Yogi and physician in New York City who specialized in rehabilitation medicine. Employed yoga. Wrote books on yoga for arthritis, back pain, and other afflictions.

JASON K. Y. FONG (1962–). Neurologist at Queen Mary Hospital, Hong Kong. Led team reporting in 1993 that a practitioner suffered a major stroke. Warned that stressful poses can cripple or kill.

MAKRAND M. GORE (1960–). Senior researcher at Kaivalyadhama, the scientific ashram of Gune south of Bombay. Studied how long advanced yogis could endure an airtight pit.

ELMER GREEN (1917–). Psychologist at the Menninger Foundation. Studied how Swami Rama used his mind alone to redirect blood flow and how students used relaxation to foster states of creative reverie.

JAGANNATH G. GUNE (1883–1966). Indian yogi and educator. Began what is considered to be the world's first major experimental study of yoga in 1924 at his ashram south of Bombay. Guided the field's development for decades.

MARSHALL HAGINS (1957–). Physical therapist at Long Island University. Participated in 2007 study that found yoga fails to meet the minimum aerobic recommendations of medical and government groups.

STEVEN H. HANUS (1954–). Physician at the Medical School of Northwestern University. Led team reporting in 1977 that a yoga practitioner suffered a major stroke after doing the Shoulder Stand.

B. K. S. IYENGAR (1918–). Yogi innovator. Studied with Krishnamacharya. Authored the 1965 book *Light on Yoga*, a global bestseller. Originated a precise style practiced around the world.

EDMUND JACOBSON (1888–1983). Physician at the University of Chicago. Taught patients how to undo muscle tension as a way to lift moods and promote healing. Authored the 1929 book *Progressive Relaxation*.

VIRGINIA E. JOHNSON (1925–). Sex researcher at Washington University in Saint Louis. Coauthored in 1966 *Human Sexual Response*. Documented long orgasms in women.

K. PATTABHI JOIS (1915–2009). Yogi innovator. Studied with Krishnamacharya. Founded style known as Ashtanga, after the eight rules of spiritual living in Patanjali's *Yoga Sutras*.

CARL JUNG (1875–1961). Swiss psychiatrist and founder of analytical psychology. Pioneered the academic study of kundalini, a yogic state characterized by strong body currents, especially up the spine. Warned in 1938 that the experience could result in madness.

SAT BIR KHALSA (1951–). Yogi and neurophysiologist at the Medical School of Harvard University. Directed many investigations of yoga, including its ability to promote sleep and reduce stage fright in musicians.

BARRY KOMISARUK (1941–). Sex researcher at Rutgers University. Studied the nature of the human orgasm and women who can think themselves into states of ecstasy.

MAIN CHARACTERS

GOPI KRISHNA (1903–1984). Kashmiri mystic. Spoke and wrote openly about his kundalini arousal. Characterized the experience as sexual in nature and a source of creativity.

TIRUMALAI KRISHNAMACHARYA (1888–1989). Guru to modern gurus. Taught yoga in Mysore, India. Trained a number of gifted students who spread modernized yoga around the globe.

WILLIAM H. MASTERS (1915–2001). Sex researcher at Washington University in Saint Louis. Coauthored the 1966 book *Human Sexual Response*. Documented fast breathing as a regular part of sexual arousal.

TIMOTHY MCCALL (1956–). Doctor and medical editor of *Yoga Journal*. Authored in 2007 *Yoga as Medicine*. Recommended that general yoga classes avoid Headstands because of the risk of injury.

RINAD MINVALEEV (1965–). Russian physiologist at Saint Petersburg State University. Led team reporting in 2004 that the Cobra position raises blood levels of testosterone, a primary sex hormone in men and women.

WILLIBALD NAGLER (1929–). Physician in Manhattan at the Weill Medical College of Cornell University. Described in 1973 a case study in which a stressful pose resulted in a major stroke.

ANDREW NEWBERG (1966–). Neuroscientist at the University of Pennsylvania Medical Center. Reported in 2009 that yoga activates the right brain—the side associated with creativity.

DEAN ORNISH (1953–). Physician known for promoting lifestyle changes to fight heart disease. Reported evidence in 2008 that yoga can lengthen cellular life spans, implying that it may fight aging.

PATANJALI (C. 400 CE). Ancient guru. Compiled *Yoga Sutras*, a collection of aphorisms about spiritual living. Urged cleanliness, good posture, breath control, ethical restraint, concentration, and meditation.

N. C. PAUL (C. 1820–1880). Indian medical doctor schooled in Calcutta. Performed what is regarded as the world's first scientific study of yoga. Authored the 1851 book *A Treatise on the Yoga Philosophy*.

LARRY PAYNE (1944–). Yoga teacher and therapist in Los Angeles. Served as founding president of the International Association of Yoga Therapists. Coauthored the book *Yoga for Dummies*.

DALE POND (1955–). Health-care specialist who investigated mystics. Helped found the Institute for Consciousness Research, a Canadian group that studied kundalini as a source of creativity.

PAUL POND (1944–). Physicist turned mystic investigator. Helped found the Institute for Consciousness Research.

JOHN P. PORCARI (1955–). Exercise physiologist at the University of Wisconsin. Participated in 2005 study that found vigorous yoga produced few aerobic benefits.

SWAMI RAMA (1925–1996). Yoga celebrity. Came to the United States from India in 1969. Underwent scientific testing in 1970 at the Menninger Foundation and displayed mental control over blood flow.

RAMAKRISHNA (1836–1886). Hindu mystic. Described the physical sensations of kundalini running up his spine.

MEL ROBIN (1934–). Yogi scientist. Worked for decades at Bell Telephone Laboratories before studying the science of yoga. Authored books in 2002 and 2009 on the inner repercussions.

W. RITCHIE RUSSELL (1903–1980). British neurologist at Oxford University. Warned in 1972 that extreme bending of the neck in strenuous yoga poses can result in debilitating strokes.

LEE SANNELLA (1916–2010). San Francisco psychiatrist. Authored a 1976 book arguing that kundalini leads to enlightenment rather than madness. Helped found the Kundalini Crisis Clinic.

BETH SHAW (1966–). Los Angeles entrepreneur. Founded YogaFit, a vigorous style that combines push-ups, sit-ups, and other repetitive exercises with yoga postures.

RANJIT SINGH (1780–1839). Maharajah of the Punjab. Sponsored a yogi's live burial in 1837, providing an early case study for the science of yoga.

SWAMI SIVANANDA (1887–1963). Guru to modern gurus. Taught Swami Vishnudevananda, who authored *The Complete Illustrated Book of Yoga* and popularized the Sivananda style.

TARA STILES (1981–). Fashion model turned yoga teacher. Opened a Manhattan studio, Strala. Authored the 2010 book *Slim Calm Sexy Yoga*.

MAIN CHARACTERS

CHRIS C. STREETER (1957–). Psychiatrist and neurologist at the Boston University School of Medicine. Led studies reporting in 2007 and 2010 that the brains of yogis show increases in a neurotransmitter that acts as an antidepressant.

JILL BOLTE TAYLOR (1959–). Neuroscientist at the Indiana University School of Medicine. Experienced right-brain euphoria after a left-brain stroke, as recounted in her 2008 book, *My Stroke of Insight.*

PATRICIA TAYLOR (1953–). Money manager turned sex therapist. Studied Tantra and authored the 2002 book *Expanded Orgasm.*

SHIRLEY TELLES (1962–). Indian physician and researcher. Led team reporting in 2011 that yoga can ease rheumatoid arthritis, a painful inflammation of the joints.

KEVIN J. TRACEY (1957–). Immunologist at North Shore University Hospital on Long Island. Reported in 2002 that the vagus nerve—a major target of yogic stimulation—wields control over the immune system.

AUREL VON TÖRÖK (1842–1912). Director of the Anthropological Museum in Budapest. Led a 1896 study of two yogis claiming to go into death-like trances.

KATIL UDUPA (1920–1992). Doctor and medical research director at the Benares Hindu University. Reported in 1974 that yoga can increase the production of testosterone, a sex hormone.

RICHARD USATINE (1956–). American physician. Helped run the family medicine program at the UCLA School of Medicine. Coauthored the 2002 book *Yoga Rx.*

AMY WEINTRAUB (1951–). Yoga teacher based in Tucson. Authored the 2004 book *Yoga for Depression.* Founded LifeForce yoga, a style designed for mood management.

CARL VON WEIZSÄCKER (1912–2007). German physicist who determined how big stars like the sun shine. Vouched for Gopi Krishna as a genuine mystic and advocated the serious study of kundalini.

DAVID GORDON WHITE (1953–). Professor of religion at the University of California, Santa Barbara. Argued that the ancient yogis sought a mental state corresponding to the bliss of sexual orgasm.

NAN WISE (1957–   ). Sex researcher at Rutgers University. Scanned the brains of women who can think themselves into ecstasy.

YOGANANDA (1893–1952). Celebrated yogi. Moved in 1920 from India to the United States. Authored the 1946 book *Autobiography of a Yogi,* the story rich in supermen and supernatural feats.

PUNJAB YOGI (C. 1837). Mystic showman. Underwent live burial in 1837 before the court of Ranjit Singh, maharajah of the Punjab. The feat became an early case study in the science of yoga.

Styles of Yoga

ANUSARA. Lighthearted. Puts emphasis on alignment of limbs and upbeat philosophy. Uses props to ease postures.

ASHTANGA. Serious. Features linked poses that flow together, as in Sun Salutation. Ties breath to postural flow. Physically demanding.

BIKRAM. Hot and sweaty. Heats practice room to loosen joints and muscles. Features twenty-six poses and two breathing exercises. Challenging.

FLOW. Graceful. Common name for styles with interconnected poses.

HATHA. Ancestral. The forerunner of all postural yoga, from medieval India. Modern forms tend to be gentle.

IYENGAR. Precise and popular. Focuses on alignment and holding poses. Uses blocks, straps, and blankets to improve positioning and avoid injury. Trains instructors for at least two years, versus weeks for many styles.

KRIPALU. Introspective. Puts emphasis on slowly introducing more challenging poses and holding them longer. Stresses awareness.

KUNDALINI. Intense. Focuses on breathing, chanting, and meditating more than postures. Seeks to awaken kundalini energy at base of spine.

POWER. Ashtanga on steroids. Many variations.

SIVANANDA. Thorough. Promotes lifestyle of moderate poses, breathing, relaxation, vegetarian diet, and cheerful attitude.

VINIYOGA. Gentle. Puts emphasis on Sun Salutations as warm-ups for more vigorous stretching.

VINYASA. Fluid. Links body movements with breath in a continuous flow. A yoga ballet.

YOGAFIT. Athletic. Targets gyms and health clubs. Mixes poses with sit-ups, push-ups, and other exercises.

Chronology

c. 2500 Clay seals of the Indus Valley civilization show figures in poses that some scholars consider the earliest known precursors of yoga. The feet of the sitting figures are tucked beneath the torso, near the genitals. The depicted individuals are seen as seeking inner heat for magic power.

438 Athens dedicates the Parthenon.

c. 400 Patanjali writes *Yoga Sutras*, a series of aphorisms on enlightenment. It describes the value of sitting comfortably for meditation but says nothing of body twists and rearrangements despite its regular citation as a founding document of postural yoga.

c. 600 Tantra emerges in India and begins to spread through Asia. It worships female deities, roots its ceremonies in human sexuality, seeks supernatural powers for material gain, and cloaks its rites in secrecy.

c. 950 Erotic sculptures of the Lakshmana temple at Khajuraho in central India depict orgies, echoing Tantric themes.

c. 1200 Gorakhnath, a Hindu ascetic of western India, fuses traditions of Tantra and body discipline into Hatha yoga. The goal is to speed enlightenment.

1288 Marco Polo visits India.

c. 1400   Swatmarama writes *Hatha Yoga Pradipika*, an early text that survives to modern times. It describes fifteen postures and many techniques of physiologic arousal.

1588   A Tantric text details a magic rite meant to let a man seduce a woman against her will.

c. 1650   The *Yoni Tantra* advises yogis to revere the female sex organ and engage in vigorous intercourse. Suggested candidates include sisters, actresses, and prostitutes.

1687   Newton posits universal gravitation and three laws of motion.

c. 1700   The *Gheranda Samhita*, a Hatha text, describes thirty-two postures and many techniques of physiologic arousal.

1772   Calcutta becomes the capital of British India.

1837   A wandering yogi undergoes live burial at the court of Ranjit Singh, the maharajah of the Punjab. The interment lasts forty days and becomes a legendary wonder.

1849   Thoreau tells a friend that he considers himself a yogi—the first known instance of a Westerner making that claim.

1851   N. C. Paul authors *A Treatise on the Yoga Philosophy*, considered the first scientific study of yoga. It seeks to explain how yogis maintain what the Indian doctor calls states of "human hibernation" and looks to yoga breathing for clues to metabolic slowdowns.

1859   Darwin authors *On the Origin of Species*.

1896   Scientists study two yogis at the Millennial Exposition in Budapest who appear to go into deathlike trances.

1918   Carl Jung treats a female client in the throes of kundalini arousal—a rush of body energy that runs from her perineum, to her uterus, to the crown of her head. His fascination with the state, central to advanced yoga, marks the beginning of Western debate on whether it results in madness or enlightenment.

1922   Gandhi is arrested for sedition during his campaign of nonco-operation with the British.

1924   Jagannath G. Gune founds an ashram south of Bombay and embarks on a major experimental study of yoga as part of a comprehensive effort to improve its image.

1926   Gune reports that the Headstand and Shoulder Stand promote blood circulation but not high pressure, casting the poses as a gentle means of physical renewal.

1927   Gune advises Gandhi on how to treat high blood pressure.

1929   Edmund Jacobson, a Chicago physician, authors *Progressive Relaxation*. It describes how easing the muscles can treat everything from headaches to depression, echoing the techniques of yoga.

1931   Gune publishes *Asanas*, the term for yoga postures. The book omits any reference to supernatural feats or Tantric rites and instead focuses on health and fitness.

1932   Kovoor T. Behanan, a Yale psychologist, arrives at Gune's ashram to study yoga.

1932   Scientists split the atom.

1933   The maharajah of Mysore in southern India hires Tirumalai Krishnamacharya to run the palace yoga studio. In time, his students become yoga's most influential gurus.

1934   The Mysore palace sends Krishnamacharya to study Gune's methods.

1937   Behanan of Yale reports on experiments in yogic breathing that produce "a retardation of mental functions."

1938   Jung calls kundalini a "deliberately induced psychotic state" that can result in "real psychosis."

1945   The first atom bomb explodes.

1946   Yogananda in *Autobiography of a Yogi* tells of mystic supermen who can read minds, see through walls, and bring the dead back to life.

1947   India becomes an independent state.

1953   Indra Devi authors *Forever Young, Forever Healthy*—the first yoga book to widely popularize the objective of ultimate health.

1957   Basu Kumar Bagchi, a scientist at the University of Michigan and onetime confidant of Yogananda, reports that yogis can achieve "an extreme slowing" of such life basics as respiration and heart rate.

1961   Bagchi reports that advanced yogis can slow but not stop their hearts—a finding that contradicts ages of miraculous claims.

1962   Scientists at Gune's ashram find that yogis in an airtight pit can withstand live burial for hours rather than weeks.

1965   B. K. S. Iyengar authors *Light on Yoga*, which becomes a global bestseller. It features the language of medicine and promotes yoga as aligned with science.

1969   Astronauts land on the moon.

1970   In a laboratory, Swami Rama demonstrates mental control over blood flowing through his palm, warming and cooling different sides.

1972   W. Ritchie Russell, a British scientist and physician, warns that pronounced bending of the neck in yoga can result in strokes and crippling disabilities.

1973   Scientists report the first of what turn out to be a number of gruesome yoga strokes.

1974   Scientists at Benares Hindu University find that yoga prompts rises in testosterone—the first evidence from a clinical laboratory that yoga can enhance sexuality.

1975   Herbert Benson, a physician at Harvard, reports that meditators can lower their respiratory rate, heart rate, blood pressure, and oxygen consumption. He calls the relaxed state *hypometabolism.*

1976   Lee Sannella, a San Francisco psychiatrist, authors a book on kundalini arousal and concludes from case studies that it promotes spiritual uplift rather than psychosis.

1976   Sannella opens the Kundalini Crisis Clinic.

1977   *Voyager 1* blasts off for Jupiter and Saturn.

1978   Scientists at Stanford University report that people in Tantric meditation undergo a variety of physiologic arousals—the reverse of yoga's usual promotion of calm serenity.

1983   Swedish scientists find that advanced yogis who breathe fast can do so without light-headedness or passing out.

1985   Czech scientists report that Tantric poses can generate surges of brain waves similar to those of lovers.

1987   The Spiritual Emergency Network finds that the typical caller to its help line is a woman with questions about kundalini.

1991   The Cold War ends.

1992   Scientists at Rutgers report that some women can think themselves into states of sexual ecstasy—an ability known clinically as *spontaneous orgasm* and popularly as *thinking off.*

1998   The National Institutes of Health begins spending public funds on yoga research, starting a wave that builds slowly in size to address such conditions as diabetes, arthritis, insomnia, depression, and chronic pain.

2001   Italian scientists report that repeating a mantra reduces respiration by about half, calming the mind.

2001   A research team at the University of California at Davis finds that yoga boosts aerobic conditioning and meets the "current recommendations to improve physical fitness and health"—a claim the sports establishment doubts and eventually seeks to disprove.

2002   Scientists at the University of British Columbia report that fast breathing can result in sexual arousal.

2002   The Consumer Product Safety Commission detects a sharp rise in yoga injuries.

2003   Yogani, an American Tantric, debuts on the Internet and draws thousands to his methods of kundalini arousal.

2004   Yogani calls kundalini a code word for sex.

2004   Russian scientists find that the Cobra position causes blood levels of testosterone to rise.

2004   Medical doctors report that fast yoga breathing ruptured a woman's lung.

2005   Analysts at the University of Virginia review seventy studies and find that yoga promotes cardiovascular health.

2006   Indian scientists report that yoga cuts the basal metabolic rate by 13 percent, threatening students with "weight gain and fat deposition." The finding contradicts a tradition of slenderizing claims.

2006   Graduates of More University in California report an experiment in which a woman stayed in an orgasmic state for eleven hours.

2007   Scientists at Columbia and Long Island universities report that vigorous yoga fails to meet the minimum aerobic recommendations of medical and government groups.

2007   A team at Boston and Harvard universities find that the brains of yoga practitioners exhibit rises in a neurotransmitter that acts as an antidepressant.

2008   A team based at the University of California at San Francisco finds that yoga increases the production of telomerase, an enzyme linked to cellular longevity.

2009   The discovery of telomerase and its role in the human body wins a Nobel Prize.

2009 Investigators at the University of Pennsylvania report that yoga can reduce hypertension and its precursors—factors linked to stroke and cardiovascular disease.

2009 Scientists in Philadelphia report that yoga activates the right brain—the side that governs creativity.

2009 A team at the University of British Columbia shows that fast breathing can heighten sexual arousal among healthy women as well as those with diminished sex drives.

2010 Analysts at the University of Maryland examine more than eighty studies and find that yoga equals or surpasses exercise in reducing stress, improving balance, diminishing fatigue, decreasing anxiety, lifting moods, and improving sleep.

2010 Indian scientists report that men and women who take up yoga enjoy wide improvements in their sex lives, including better desire, arousal, satisfaction, and emotional closeness with partners.

2011 Physicians in Taiwan find that yoga lessens the incidence of spinal deterioration.

2011 Indian scientists report that yoga can ease trauma from rheumatoid arthritis, a painful inflammation of the joints that afflicts millions of people.

2011 Connecticut researchers find that elderly women who take up yoga improve their sense of balance.

THE SCIENCE

OF YOGA

Prologue

Yoga is everywhere among the affluent and the educated. The bending, stretching, and deep breathing have become a kind of oxygen for the modern soul, as a tour of the neighborhood shows rather quickly. New condo developments feature yoga studios as perks. Cruise ships tout the accomplishments of their yoga instructors, as do tropical resorts. Senior centers and children's museums offer the stretching as a fringe benefit—*Hey, parents, fitness can be fun.* Hollywood stars and professional athletes swear by it. Doctors prescribe it for natural healing. Hospitals run beginner classes, as do many high schools and colleges. Clinical psychologists urge patients to try yoga for depression. Pregnant women do it (very carefully) as a form of prenatal care. The organizers of writing and painting workshops have their pupils do yoga to stir the creative spirit. So do acting schools. Musicians use it to calm down before going on stage.

Not to mention all the regular classes. In New York City, where I work, it seems like a yoga studio is doing business every few blocks. You can also take classes in Des Moines and Dushanbe, Tajikistan.

Once an esoteric practice of the few, yoga has transformed itself into a global phenomenon as well as a universal icon of serenity, one that resonates deeply with tense urbanites. In 2010, the city of Cambridge, Massachusetts, began illustrating its parking tickets with a series of calming yoga poses.

The popularity of yoga arises not only because of its talent for undoing stress but because its traditions make an engaging counterpoint to modern life. It's unplugged and natural, old and centered—a kind of anti-civilization pill that can neutralize the dissipating influence of the Internet and the flood of information we all face. Its ancient serenity offers a new kind of solace.

An indication of yoga's social ascendency is how its large centers often get housed in former churches, monasteries, and seminaries, the settings frequently rural and inspirational. Kripalu, on more than three hundred rolling acres of the Berkshires in western Massachusetts, was once a Jesuit seminary. Each year its yoga school graduates hundreds of new teachers. And they in turn produce thousands of new yogis and yoginis, or female yogis.

Even the White House is into yoga. Michelle Obama made it part of Let's Move—her national program of exercise for children, which seeks to fight obesity. The First Lady talks about yoga on school visits and highlights the discipline at the annual Easter Egg Roll, the largest public event on the White House social calendar. Starting in 2009, the egg roll has repeatedly featured a Yoga Garden with colorful mats and helpful teachers. The sessions start early and go throughout the day.

On the White House lawn in 2010, an adult dressed as the Cat in the Hat—a character from the Dr. Seuss book—did a standing posture on one leg. A tougher demonstration featured five yogis simultaneously upending themselves in Headstands. At the 2011 event, the Easter Bunny did a tricky balancing pose. The children watched, played along, and took home a clear message about what the President and First Lady considered to be a smart way of getting in shape.

Yoga is one of the world's fastest-growing health and fitness activities. The Yoga Health Foundation, based in California, puts the current number of practitioners in the United States at twenty million and around the globe at more than two hundred and fifty million. Many more people, it says, are interested in trying yoga. To spread the word, the foundation organizes Yoga Month—a celebration every September that blankets the United States with free yoga classes, activities, and health fairs.

By any measure, the activity is too widespread and its participants too affluent for advertisers and the news media to ignore. Health and beauty magazines do regular features. *The New York Times*, where I work, has run hundreds of articles and in 2010 began a regular column, Stretch. It has profiled everything from studios that offer hot yoga in overheated rooms to a gathering of thousands in Central Park that its organizers called the largest yoga class on record. A main attraction of that event was the corporate gifts. Participants got JetBlue yoga mats, SmartWater bottles, and ChicoBags filled with giveaways. The allure was so great that many people got stuck in entrance lines before a downpour chased everybody away.

Yoga may be in the air culturally. But it is also quite visibly a big business. Merchants sell mats, clothes, magazines, books, videos, travel junkets, creams, healing potions, shoes, soy snacks, and many accessories deemed vital to practice—as well as classes. Purists call it the yoga industrial complex. Increasingly, the big financial stakes have upended the traditional ethos. Bikram Choudhury, the founder of Bikram Yoga, a hot style, copyrighted his sequence of yoga poses and had his lawyers send out hundreds of threatening notices that charged small studios with violations. He is not alone. In the United States, yoga entrepreneurs have sought to enhance their exclusivity by registering thousands of patents, trademarks, and copyrights.

Market analysts identify yoga as part of a demographic known as LOHAS—for Lifestyles of Health and Sustainability. Its upscale, well-educated individuals are drawn to sustainable living and ecological initiatives. They drive hybrid cars, buy natural products, and seek healthy lifestyles. Yoga moms (a demographic successor to soccer moms) are an example. According to marketing studies, they tend to buy clothes for their children from such places as Mama's Earth, its goods made from organic cotton, hemp, and recycled materials.

One factor that distinguishes modern yoga from its predecessors is its transformation from a calling into a premium lifestyle. Another is that women make up the vast majority of its practitioners, a fact that dramatically influences the nature of its marketplace. Women buy more books than men, read more, spend more on consumer goods, and pay more attention to their health and appearance.

Yoga Journal—the field's leading magazine, founded in 1975—claims two million readers and identifies its audience as 87 percent women. It revels in their quality, citing high incomes, impressive jobs, and good educations. A brochure for prospective advertisers notes that more than 90 percent have gone to college.

The colorful pages of the magazine offer a vivid example of how companies target the demographic. Hundreds of ads promote skin-care products, sandals, jewelry, natural soaps, special vitamins and enzymes, alternative cures and therapies, smiling gurus, and ecofriendly cars. Each issue features an index to advertisers. One of my favorites is Hard Tail, a clothing line whose ads feature attractive women in striking poses. "Forever," reads the minimalist copy.

Another is Lululemon Athletica, a hip brand of yoga clothing known for its form-fitting apparel, most especially its ability to shape and display the buttocks to best advantage. Recently, a market analyst identified Lulu's signature item as the $98 Groove Pant, "cut with all kinds of special gussets and flat seams to create a snug gluteal enclosure of almost perfect globularity, like a drop of water."

All of which bears on what yoga (as opposed to its accessories) does for the body and mind or, more precisely, on what gurus, spas, books, instructional videos, merchants, television shows, magazines, resorts, and health clubs say that it does.

In this regard, it is important to remember that yoga has no governing body. There's no hierarchy of officials or organizations meant to ensure purity and adherence to agreed-upon sets of facts and poses, rules and procedures, outcomes and benefits. It's not like a religion or modern medicine, where rigorous schooling, licensing, and boards seek to produce a high degree of conformity. And forget about government oversight. There's no body such as the Consumer Product Safety Commission or the Food and Drug Administration to ensure that yoga lives up to its promises. Instead, it's a free-for-all—and always has been. Over the ages, that freedom has resulted in a din of conflicting claims.

"The beginner," notes I. K. Taimni, an Indian scholar, "is likely to feel repulsed by the confusion and exaggerated statements." Taimni wrote that a half century ago. Today the situation is worse. For one thing, the explosion in publishing—print and electronic—has amplified the din into a cacophony. Another factor is the profit motive.

Billions of dollars are now at stake in public representations of what yoga can do, and the temptations are plentiful to lace declarations with everything from self-deception and happy imprecision to willful misrepresentations and shadings of the truth. Another temptation is to avoid any mention of damage or adverse consequences—a silence often rooted in economic rationalizations. Why tell the whole story if full disclosure might drive away customers? Why limit the sales appeal? Why not let the discipline be all things to all people?

Anyone who has done yoga for a while can rattle off a list of benefits. It calms and relaxes, eases and renews, energizes and strengthens. It somehow makes us feel better.

But beyond such basics lies a frothy hodgepodge of public claims and assurances, sales pitches and New Age promises. The topics include some of life's most central aspirations—health, attractiveness, fitness, healing, sleep, safety, longevity, peace, willpower, control of body weight, happiness, love, knowledge, sexual satisfaction, personal growth, fulfillment, and the far boundaries of what it means to be human, not to mention enlightenment.

This book cuts through the confusion that surrounds modern yoga and describes what science tells us. It unravels more than a century's worth of research to discern what's real and what's not, what helps and what hurts—and nearly as important, why. It casts light on yoga's hidden workings as well as the disconcerting reality of false claims and dangerous omissions. At heart, it illuminates the risks and the rewards.

Many, it turns out, are unfamiliar.

I came to this book as a knowledgeable amateur. During my freshman year of college, in 1970, I got hooked on yoga because it felt good and seemed to make me healthier in body and mind. My first teacher said it was important to do some—even a little—every day. That's always been my goal, despite the usual struggle with good intentions. Yoga has become a good friend to whom I turn no matter how crazy my life gets.

I began my research in 2006. My plan was simple. I'd track down the best science I could find and answer a lot of questions that I had accumulated over the decades, things I had wondered about but never had a chance to explore.

My first surprise was how yoga had morphed into a confusing array of styles and brands. I knew enough to understand that the origin of it all was Hatha yoga—the variety that centers on postures, breathing, and drills meant to strengthen the body and the mind (as opposed to the yogas of ethics and religious philosophy). Today, Hatha and its offspring are the most widely practiced forms of yoga on the planet, having produced scores of variations that range from local styles in most every country to such ubiquitous global brands as Iyengar and Ashtanga.

My enthusiasm for gyms and swimming also gave me a reasonable perspective on how yoga differs from regular exercise. In general (with exceptions we'll study closely), it goes slow rather than fast, emphasizing static postures and fluid motions rather than the rapid, forceful repetitions of, say, spinning or running. Its low-impact nature puts less strain on the body than traditional sports, increasing its appeal for young people as well as aging boomers. In terms of physiology, it takes a minimalist approach to burning calories, contracting muscles, and stressing the body's cardiovascular system. Perhaps most distinctively, it places great emphasis on controlling the breath and fostering an inner awareness of body position. Advanced yoga, in turn, goes further to encourage concentration on subtle energy flows. Overall, compared to sports and other forms of Western exercise, yoga draws the attention inward.

I began examining the yogic literature with a sense of wariness. Long ago, while working at the University of Wisconsin on a study of respiratory physiology, I came across a flat contradiction to one of modern yoga's central tenets—that fast breathing floods the body and brain with revitalizing oxygen. In contrast, a textbook I was reading at the time said the pace of human respiration "can drop to one-half or rise to over one hundred times normal without appreciably influencing the amount of blood oxygenated." I see now that, in 1975, I underlined that passage quite heavily.

Unfortunately, my survey lived up to my low expectations. Some books and authors shone brightly. (See Further Reading for a list.) But on the whole, I found the literature dull with dreaminess, assertions with no references, and a surprising number of obvious untruths. I wanted tips for tracking down good science but instead got a muddle. The writing, old and new, turned out to run toward the curiously dogmatic and, at best, to contain only a smattering of science. Much of it was similar to what Richard Feynman, a founder of modern physics, disparaged as cargo-cult science—that is, material that appears scientific but lacks factual integrity.

By contrast, my plunge into the scientific literature left me heartened. Federal officials at the National Institutes of Health in Bethesda, Maryland, run a wonderful electronic library of global medical reports known as PubMed, short for Public Medicine. It showed that scientists had written nearly one thousand papers related to yoga—the number rising in the early 1970s and soaring in recent years, with reports added every few days. The studies ranged from the flaky and superficial to the probing and rigorous. The authors included researchers at Princeton and Duke,

Harvard and Columbia. Moreover, the field had gone global. Scientists in Sweden and Hong Kong were publishing serious papers.

But the closer I looked, the more I judged this body of information to be rather limited. Some general topics had been covered fairly well—for instance, how yoga can relax and heal. But many others were ignored, and much of the published science turned out to be superficial. For instance, studies for the approval of a new drug can require the participation of hundreds or even thousands of human subjects, the large numbers increasing the reliability of the findings. In contrast, many yoga investigations had fewer than a dozen participants. Some featured just one individual.

The superficiality turned out to have fairly obvious roots. Research on yoga was often a hobby or a sideline. It had no big corporate sponsors (there being no hope of discoveries that could lead to expensive pills or medical devices) and relatively little financial support from governments. Federal centers tend to specialize in advanced kinds of esoteric research as well as pressing issues of public health, with their investigations typically carried out at institutes and universities. In short, modern science seemed to care little.

The exception turned out to be areas where yoga intersected other disciplines or made bold claims strongly at odds with the conventional wisdom. Such crossroads proved to be scientifically rich. For instance, scientists interested in sports medicine and exercise physiology had lavished attention on yoga's fitness claims. So, too, physicians had zeroed in on yoga's reputation for safety.

The limitations of the current literature sent me casting a wide net, and I immediately made a big catch. It was a very old book—*A Treatise on the Yoga Philosophy*—written by a young Indian doctor and published in 1851 in Benares (now known as Varanasi), the ancient city on the Ganges that marks the spiritual heart of Hinduism. It came to my attention because a few Western scholars had referred to it in passing.

I got lucky and found that Google Books had recently scanned Harvard's copy into its electronic archive, so I was able to download the whole thing in a flash. Its language was archaic. But the author had addressed the science of yoga with great skill, illuminating an important aspect of respiratory physiology that many authorities still get wrong today.

The book surprised me because I had been told that Indian research on yoga—though pioneering—was typically of poor quality. But I kept finding gems. A curious scientist, often working in India or born in India and doing research abroad, would address some riddle of yoga and make important finds. It happened not only with respiratory physiology but psychology, cardiology, endocrinology, and neurology. The scientists often acted with rigor, going against the day's tide.

Intrigued, I traveled to India to learn more about these early investigators and eventually came to see them as a kind of intellectual vanguard. Their reports tended to predate the electronic archives of PubMed, making them all but invisible to modern researchers. But their findings turned out to be central to the field's development.

As I widened my research, I had the great good fortune to sit at the feet of Mel Robin, a veteran yoga teacher and star of yoga science. Mel had worked at Bell Telephone Laboratories (the birthplace of the transistor, the heart of computer chips) for nearly three decades before turning to an investigation of yoga. His labor of love produced two massive books totaling nearly two thousand pages. What Mel did uniquely was roam far beyond the literature of yoga to show how the general discoveries of modern science bear on the discipline. His example encouraged the kind of independent thinking I had begun at the University of Wisconsin.

Over the years, the widening of my research brought me into contact not only with Mel but a wonderful variety of scientists, healers, yogis, medical doctors, mystics, federal officials, and other students of what science tells us about yoga. If science is the spine of this book, they are the flesh and blood.

My focus is practical. In places, the book touches on topics of Eastern spirituality—meditation and mindfulness, liberation and enlightenment—but makes no effort to explore them. Rather, it zeroes in relentlessly on what science tells us about postural yoga. I mean no disrespect to the Hindu religion or spiritual traditions that embrace the big picture. But if this book succeeds, it does so because it limits itself to a poorly known body of reductionist findings. Even so, I should note that I view the scientific process as limited and unable to answer the most important questions in life, as does any true believer. The epilogue explores what may lie beyond.

In the end, my examination revealed not only a wealth of findings but a remarkable lack of knowledge among yogis, gurus, and practitioners about the reports and investigations. This is pure speculation. But I'd be surprised if the community knows a hundredth or even a thousandth of what scientists have learned over a century and a half.

This book tells that story. In essence, it offers an impartial evaluation of an important social phenomenon that began to stir millennia ago. And if I may, it is the first to do so.

I have structured this book to start with issues of common interest and to end with topics that are less familiar. That flow, it turns out, parallels the development of scientific interest over the decades. So the book has a loose chronological organization.

The portrait of yoga that emerges is quite different in important respects from the usual claims. In some cases, the news is better.

For instance, a number of teachers credit yoga with powers of sexual renewal. The science not only confirms that claim but shows how specific poses can act as aphrodisiacs that produce surges of sex hormones and brain waves indistinguishable from those of lovers. More generally, recent clinical studies give substance to the idea that yoga can improve the sex lives of men and women, documenting how new practitioners report not only enhanced feelings of pleasure and satisfaction but emotional closeness with partners.

The health benefits also turn out to be considerable. While many gurus and how-to books praise yoga as a path to ultimate well-being, their descriptions are typically vague. Science nails the issue.

For example, recent studies indicate that yoga releases natural substances in the brain that act as strong antidepressants, suggesting great promise for the enhancement of personal health. Globally, depression cripples more than one hundred million people. Every year, its hopelessness results in nearly a million suicides.

Amy Weintraub, a major figure in this book, recounts how yoga saved her life by cutting through clouds of despondency.

But if some findings uplift, others contradict the onslaught of bold claims and proffered cures.

Take body weight—a topic of enormous sensitivity for anyone trying to look good. For decades, teachers of yoga have hailed the discipline as a great way to shed pounds. But it turns out that yoga works so well at reducing the body's metabolic rate that—all things being equal—people who take up the practice will burn fewer calories, prompting them to gain weight and deposit new layers of fat. And for better or worse, scientists have found that the individuals most skilled at lowering their metabolisms are women. Of course, other aspects of yoga *do* fight pounds successfully. The discipline builds body awareness and its calming influence can help reduce stress eating. Most yoga teachers are lithe, not lumpy. But when yoga succeeds at weight control, the scientific evidence suggests that it does so in spite of—not because of—its basic impact on the human metabolism.

That's one of yoga's dirty little secrets. It turns out there are plenty of others, some quite significant.

Yoga has produced waves of injuries. Take strokes, which arise when clogged vessels divert blood from the brain. Doctors have found that certain poses can result in brain damage that turns practitioners into cripples with drooping eyelids and unresponsive limbs.

Darker still, some authorities warn of madness. As Carl Jung put it, advanced yoga can "let loose a flood of sufferings of which no sane person ever dreamed." Many yoga books cite Jung approvingly but always seem to miss that quote. Even so, it represented his considered opinion after two decades of study and reflection.

Overall, the risks and benefits turned out to be far greater than anything I ever imagined. Yoga can kill and maim—or save your life and make you feel like a god. That's quite a range. In comparison, it makes most other sports and exercises seem like child's play.

My research has prompted me to change my own routine. I have deemphasized or dropped certain poses, added others, and in general now handle yoga with much greater care. I hope you benefit, too.

I see this book as similar to informed consent—the information that the subjects of medical experiments and novel treatments are given to make sure they understand the stakes, pro and con.

To me, the benefits unquestionably outweigh the risks. The discipline on balance does more good than harm. Still, yoga makes sense only if done intelligently so as to limit the degree of personal danger. I'm convinced that even modest precautions will avert waves of pain, remorse, grief, and disability.

• • •

The heroes of this book are the hundreds of scientists and physicians who toiled inconspicuously over the decades to uncover the truth despite the obstacles of scarce funding and institutional apathy. Their early inquiries not only began the process of illuminating yoga but, as it turns out, produced a remarkable side effect. They helped transform the nature of the discipline.

Yoga at the start was an obscure cult steeped in magic and eroticism. At the end, it fixated on health and fitness.

To my surprise, it turned out that science played an important role in the modernization. As investigators began to show how the ostensible wonders of yoga had natural explanations, the discipline worked hard to reinvent itself. A new generation of gurus downplayed the rapturous and the miraculous for a focus on material well-being. In essence, they turned yoga on its head by elevating the physical over the spiritual, helping create the secular discipline now practiced around the globe.

The first chapter details this upheaval. The tale is important not only for revealing the origins of the health agenda but for introducing main characters and themes. For instance, it turns out that a number of yoga miracles—if demonstrably untrue—nonetheless involve major alterations of physiology that can produce a wealth of real benefits. They can lift moods. They can fight heart disease. The newest research indicates that they may even slow the body's biological clock.

Not that science has all the answers.

To the contrary, the investigation of the discipline began in response to an astonishing spectacle nearly two centuries ago that still poses a number of fundamental questions today.

The science of yoga does more than reveal secrets. It can also shed light on real mysteries.

I

HEALTH

Ranjit Singh was an ugly little man who liked to surround himself with beautiful women. In childhood, smallpox had taken his left eye and pitted his face. He was unlettered. But Singh built an empire through force of character, uniting the warring tribes of western India. He became maharajah of the Punjab and amassed great wealth, including the Koh-i-Noor, at the time the world's largest diamond. He could be generous. Though a Sikh, he gave a Hindu temple a ton of gold. Singh was a military genius and a humane despot. Most of all, he knew men.

In 1837, Singh learned that a wandering yogi had approached the court to propose live burial as a demonstration of his spiritual powers. The king agreed to sponsor the entombment but undertook a number of precautions. The holy man would be interred in a small building near the palace. In preparation, Singh had three of its four doorways sealed with bricks and mortar, turning the open structure into something resembling a jail—or, less optimistically, a crypt.

Military officers, as well as European doctors, watched as the yogi arranged himself into a sitting posture. It was most likely a Full Lotus, with legs crisscrossed and feet atop the thighs. One observer likened the pose to that of "a Hindoo idol." Attendants then wrapped the yogi in white linen and placed him inside a wooden box. It rested in a shallow pit below the building's floor. No dirt was applied because the yogi had expressed concern about ants attacking his body. The maharajah's men did, however, secure the box with lock and key. They then padlocked the door at the building's entrance and erected a mud wall to seal off the improvised cell from the outside world.

The building was judged to have no hole that could admit air, and no passageway through which food could pass. Sentries kept watch day and night. A senior officer of the court came by periodically to check on security and report back to the maharajah.

The interment lasted forty days and forty nights—a period that, from biblical times, has stood for completeness and unbroken cycles. Then the king rode up on an elephant, dismounted before his assembled court, and surveyed the results.

The linen bag looked mildewed, as if it had lain undisturbed for a long time. The yogi's legs and arms proved to be cold, stiff, and shriveled, his skin pale. No pulse could be detected.

Then his eyes opened.

The yogi's body convulsed violently. His nostrils flared. A faint heartbeat could now be heard. After a few minutes, his eyes dilated. His color returned.

Seeing the king nearby, the yogi asked in a low, barely audible voice, "Do you believe me now?"

Yoga in centuries past was a mystic wonderland in which the practices differed from our own in ways that ranged from the mundane to the almost unimaginable. Take instruction. It was done in private rather than in classes. More important, relatively few women did yoga. That was understandable given the chauvinistic leanings of old societies. The most radical difference centered on the lifestyles of the men.

Yogis were often vagabonds who engaged in ritual sex or showmen who contorted their bodies to win alms—even while dedicating their lives to high spirituality. The Punjab yogi was no exception. Chroniclers report that he always did his burial feats "for good compensation," as one put it. After surviving his forty-day interment, he was presented with a pearl necklace, gold bracelets, pieces of silk, and shawls of a kind "usually conferred by the Princes of India on persons of distinction."

Yogis were as much gypsies as circus performers. They read palms, interpreted dreams, and sold charms. The more pious often sat naked—their beards uncut and hair matted—and smeared themselves with ashes from funeral pyres to emphasize the body's temporality.

The Kanphata yogis, a large sect, had reputations as child snatchers. To obtain new members, they would adopt orphans and, when the opportunity arose, buy or steal children. Understandably, good families dreaded their presence. At times, bands of yogis would prey on trade caravans and descend on merchants to extort food and money. When hired as guards, violent orders formed what we would now call protection rackets.

Some yogis smoked ganja and ate opium. Some carried begging bowls. A few were surely saints. But British officialdom as well as educated Indians came to resent the holy men as not only potentially dangerous but as economic drains on society. A British census summed up the condescension tartly by putting yogis under the heading "miscellaneous and disreputable vagrants."

No small part of the disrepute centered on sex. Spiritually, the objective of the yogi was to achieve a blissful state of consciousness in which the male and female aspects of the universe merged into a realization of oneness. That union (the word "yoga" means union) resulted in enlightenment. But a main path was sexual ecstasy—a veiled part of the agenda that modern research has recently uncovered. David Gordon White, one of the field's preeminent scholars, who teaches at the University of California, Santa Barbara, noted in a 2006 book that the ancient yogis sought a divine state of consciousness "homologous to the bliss experienced in sexual orgasm."

The path to the ecstatic union was known as Tantra. Hugely popular, it rejected the caste system, pulled in converts by the cartload, and gave rise to religious authorities who wrote thousands of texts and commentaries. It reveled in magic, sorcery, divination, ritual worship (especially of goddesses), cultic rites of passage, and sacred sexuality.

In the West, Tantra is best known as an originator of sexual rites. And rites there were—enough to raise protests from the Hindu and Buddhist orthodoxy of the day. The main charge was that Tantrics indulged in sexual debauchery under the pretext of spirituality.

So too the Punjab yogi, a good Tantric. As his reputation rose, his behavior became so bad that the maharajah considered throwing him out of the kingdom. But the yogi left of his own accord. He did so with spirits high, eloping to the mountains with a young married woman.

Over the centuries, Tantra underwent various degradations that reached their nadir with the Aghori—a cannibal sect that ate the flesh of human corpses, drank urine and liquor from human skulls, lived in cremation grounds and dunghills, and reviled all social convention, supposedly to court public disapproval as tests of humility. The primal ascetics also practiced ritual cruelty and seasonal orgies. Scholars of religion tend to avoid the gory details but do mention such things as an Aghori predilection for incest. In any event, the worst behaviors associated with Tantra were so extreme that the overall practice came to be condemned as a threat to society.

Another way that old yoga differed from our own was its formulation of Hatha—or postural yoga. The principles were laid out in the *Hatha Yoga Pradipika*. The holy book of the fifteenth century represents the discipline's earliest extant text.

The book lavished attention on body parts that have nothing in common with the modern focus, including the penis, vagina, scrotum, and anus. Over and over, it recommended sitting postures meant to exert pressure on the perineum—the area between the anus and genitals that is sensitive to erotic stimulation. "Press the perineum with the heel of the foot," the text advised. "It opens the doors of liberation."

Today, the term of art for a yoga posture is *asana*. But the word in Sanskrit actually means "seat"—harkening back more than a millennium to the days when postural yoga referred to nothing more complicated than sitting in a relaxed position for meditation. The *Hatha Yoga Pradipika* put bold new emphasis on sitting postures and stimulating acts. It said nothing of standing poses or the kinds of fluid movements so popular in contemporary yoga classes.

The book also told how to extend the duration of lovemaking—and focused its advice on males, reflecting yoga's ancient bias. It called for "a female partner" but conceded that a willing consort was something "not everyone can obtain."

One instruction claimed that a particular technique would produce such steely control in sexual relations that the yogi would release no semen even if "embraced by a passionate woman." The goal was to slowly raise the levels of excitement, the couple approaching but never quite reaching orgasm, their ecstasy going on and on, the two becoming one, transcending all opposites.

If such depictions of Hatha yoga strike the modern reader as bizarre, it is because contemporary books and teachers seldom refer to the origins of the practice. But in truth, *Hatha is a branch of Tantra*. It was developed as a way to speed the Tantric agenda, to make enlightenment happen by the precise application of willpower and the redirection of libidinal energy rather than by some nebulous mix of piety and contemplation. The Sanskrit root of Hatha is *hath*—"to treat with violence," as in binding someone to a post, according to the Monier-Williams Sanskrit dictionary, by an Oxford professor. So, too, Hatha means violence or force. The discipline arose in a carefully structured campaign of vigorous activity meant to promote the quick attainment of enlightenment through ecstasy.

So it is that a number of scholars translate Hatha yoga as "violent union." Other specialists render it as "union from violence or force" to put the emphasis on the illumination rather than its means of attainment. In either case, such definitions seldom—if ever—show up in the popular literature. The New Age approach is to embrace the poetry of Sanskrit and divide Hatha into *ha* and *tha,* for sun and moon. That interpretation casts the word itself as an esoteric uniting of opposites and typically omits any reference to force or violence.

A final way that old yoga differed from our own was its emphasis on the miraculous. For ages, the sacred literature of India had portrayed yogis as able to fly, levitate, stop their hearts, suspend their breathing, vanish, walk through walls, project themselves into other bodies, touch the moon, survive live burial, make themselves invisible, die at will, walk on water, and—like Jesus of Nazareth—bring the dead back to life. They were hailed as miracle workers. Their unusual abilities had a name—*siddhis*. The Sanskrit word means success or perfection and is a yogic term of art for the otherworldly powers. Patanjali, the Indian sage who laid out the fundamentals of mystic yoga some sixteen centuries ago, devoted an entire chapter of his aphorisms to the otherworldly feats, including such talents as reading minds and predicting the future.

Astonishing claims filled the pages of *Hatha Yoga Pradipika.* It said practitioners could neutralize poisons, destroy all diseases, annihilate old age, obviate evil, and achieve immortality—not to mention doing away with constipation, wrinkles, and gray hair.

Yogi warriors made miraculous claims to enhance their battlefield image, according to William Pinch, a scholar at Wesleyan University. Yoga, he said, conferred a reputation of invincible power. "There was a clear tactical advantage of believing, and having your enemy believe, that you were immortal."

The basic accomplishment that bestowed the gift of the miraculous on the lowly practitioner was the attainment of samadhi—the state of transcendent bliss in which the yogi became one with the universe. The adept did so after learning how to move all the currents of prana, the body's energy, up the spine into the head. At that point, according to *Hatha Yoga Pradipika,* the yogi became "as if dead."

Some yogis entered the euphoric state for the purposes of spiritual enlightenment. Others—like the Punjab yogi, true to the diversity of the Tantric brotherhood—did so for entertainment and profit.

The dramatic success of the live burial astonished many people—and not just at the court of Ranjit Singh. Books describing the feat were published in Vienna, London, and New York. The educated world marveled at the accomplishment and wondered at its explanation. Claude M. Wade, the British liaison to the maharajah's court and an eyewitness to the yogi's exhumation, cautioned his peers that it would be "presumptuous to deny to the Hindoos the possible discovery or attainment of an art which has hitherto escaped the researches of European science."

At the time of the burial, N. C. Paul was entering medical school in Calcutta (today known as Kolkata) and, as a new scientist, paid close attention. After all, the spectacle appeared to defy the laws of nature. His curiosity led him to write a book—*A Treatise on the Yoga Philosophy.*

It featured the live burial and, as it turned out, marked the birth of a new science.

Who was Paul? No scholar or book gave him more than a passing reference. I knew little until I went to Calcutta, a city crackling with energy despite the monsoon heat.

Blaring horns and bad traffic greeted my cab ride to his medical school—a place I expected to bear the tidy imprint of its British founders. Instead, it was bedlam. Stray dogs, sick people, and students roamed a warren of broken buildings and fallen trees. Walls bore faded posters. I grew apprehensive as I neared the library, increasingly uneasy but still eager to learn about the world's first scientist who sought to free yoga from its mythic past.

I climbed a circular stairway past loose wires, cobwebs, and pieces of shattered concrete. The library had a high ceiling and dark wood that bespoke past elegance. But decay had set in. Cabinets with glass doors held row upon row of old books—the dust inside so thick that it obscured the titles. With a start, I realized that the cases had become mausoleums. Overhead, gauzy spider webs hung down like props in a horror movie. The smiling librarian in her colorful sari seemed slightly embarrassed— but not too much. It turned out that she, like everyone else, knew nothing of Paul and little of the school's early days. The Bengal Medical College had been founded in 1835 as the first school of European medicine in Asia, its red brick and white trim meant to symbolize a new era.

In a panic, I sped across town to the National Library, a colonial relic on a lush campus. For days I pored through old books and reports. Nothing. Not a trace. Some records were so fragile that they fell apart in my hands. Worms had eaten their way through many books, leaving trails of missing letters and words. I made little piles of debris at my desk. Book after book. Nothing.

Finally, in the last volume, there it was—a life sketch of Paul. He was included in a list of graduates from the Bengal school who had gone into the colonial medical service. Dates. His first job. What he earned. A break in my research after days of nervous sweat.

My luck increased when I met P. Thankappan Nair, a short man of seventy-four who looked like Gandhi. He had written dozens of books on Calcutta and proved to be a treasure trove of ideas, kindness, energy, and common sense. Nair made journalism seem respectable.

We visited historians, archives, literary societies, and more, traveling by bus, subway, bicycle rickshaw, and train (open doors, looking out over villages and smoky morning fires). He refused money. Nair explained that he did such things out of a sense of civic duty.

Paul was a native Bengali (his given name was Nobin Chunder Pal or, in some iterations, Navina Chandra Pala or Nobin Chundra Pal) who had climbed the social ladder by virtue of a good education. The British rulers of early nineteenth-century India exploited the nation ruthlessly. But they also established schools for native youth in which the curriculum was European and the language of instruction was English. The idea was to build a class of skilled underlings to aid the empire's administration.

Paul represented one of the early successes. In Calcutta, the first capital of British India, he enrolled in medical school and applied himself to the city's intellectual life. At the Society for the Acquisition of General

Knowledge, a hotspot of upward mobility, he heard lectures on such racy topics as "The Interests of the Female Sex."

Paul graduated in 1841 and proudly displayed behind his name the initials G.B.M.C.—Graduate of Bengal Medical College. It announced his elite status, as did the Europeanizing of his surname.

His big break came when he was transferred to Benares. For Hindus, it was the holiest city in India, located on the banks of the Ganges, the most sacred of rivers. His post gave him a commanding view of yogic life. Hundreds of temples lined the river, and pilgrims came from all over to bathe, or to cremate the bodies of loved ones, eager to wash away sins and win salvation. So, too, mystics flocked to the ghats, or wide stone steps, to take purifying dips in the river, to practice yoga, and to meditate. At least one slept on a bed of nails. The Buddha gave his first sermon nearby. For ages, Benares served as the heart of Hinduism, playing the same kind of role that Mecca does for Muslims and the Vatican for Roman Catholics. Many Hindus still consider it the holiest place on earth.

Paul's *Treatise on the Yoga Philosophy* appeared in 1851. London that year was holding its Great Exhibition. It was meant to draw attention to Britain as the leader of the industrialized world.

The regimental surgeon seemed eager to show that the colonies, too, could participate in the march of progress.

Paul commented on such things as the Aghori, the cannibal sect, noting the group's heavy consumption of liquor. But his book generally ignored the remarkable diversity of Hindu mystics and instead zeroed in on their talent for life suspension or, as he put it in his preface, "abstaining from eating and breathing for a long time, and of becoming insensible to all external impressions."

His principal case study was the forty-day burial. The Punjab yogi's triumph over death, Paul noted, "has puzzled a great many learned men of Europe." But he—lowly Bengali surgeon that he was—would deign to enlighten his peers.

Paul's explanation had nothing to do with challenging poses or purifications. Nor did it deal with anatomy. Instead, he focused on an unseen factor that scientists of his day were busy measuring and evaluating in people and the environment.

It was carbon dioxide, the waste product of cellular respiration that we all exhale. He had learned much about the basics of the transparent gas in medical school and quickly realized that yogic rituals worked to bottle it up inside the body. The main technique of manipulation was pranayama—the Sanskrit term for breathing exercises and, more literally, control of the life force. *Prana* means "vital force" and *yama* means "to restrain or control."

Paul gave many examples of how yogis manipulate their breathing to discharge less. For instance, he described a common practice in which yogis take fewer breaths. Such retention, he wrote, "has a remarkable effect" on reducing how much carbon dioxide a person exhales.

His figures showed big drops over an extraordinary range of breathing rates, starting with the fastest. A relaxed person takes roughly fifteen breaths per minute. Paul's findings showed that a yogi who took ninety-six breaths a minute would expel seventy-nine cubic inches of carbon dioxide. At twenty-four breaths, the level fell by more than half to twenty-four cubic inches. At six breaths—far below the pace of normal breathing—the flow dropped again by more than half to ten cubic inches. Three breaths per minute cut the level still further to five cubic inches. And one breath per minute took the measurement down to two cubic inches. In other words, slow breathing produced huge drops in the exhalation of carbon dioxide.

Over a dozen pages, Paul told how many other yogic practices worked similarly to lower the outflow. For instance, he said the repetition of "om," the holy syllable of yoga, "materially diminished" the carbon loss. Another tactic was to simply rebreathe the same air—a move Paul called "one of the easiest methods" for entering a euphoric trance.

The culmination of Paul's analysis centered on a common feature of yogic life that worked inconspicuously to concentrate stale air for rebreathing. It was to live in a gupha, or cave, a kind of subterranean retreat with little light or ventilation, made for long periods of contemplative bliss. Paul noted that a yogi's assistant would block the small doorway of the gupha with clay—not unlike the mud that was used to seal the Punjab yogi's crypt. That produced "a confined atmosphere," Paul noted, where the yogi expired less carbon dioxide than would be the case "in the free ventilated air."

In a bold display of naturalism, Paul laid out a metaphor that avoided any hint of the miraculous and replaced it with a common feature of the natural world. The life suspensions of the yogis, he proposed, were simply cases of human hibernation, and the gupha of the saints were analogous to the burrows of hibernating animals.

Paul drove his point home by observing that hibernation was a common strategy that many different animals used to conserve energy. So the holy men resembled bats, hedgehogs, marmots, hamsters, and dormice, he argued. So too, like hibernating animals, yogis lined their caves with such insulating materials as grass, cotton, and sheepswool. "The guphá is as indispensably necessary to the Yogi," Paul wrote, as a burrow is to "the hibernating animals."

It was human hibernation, he concluded, that let the Punjab yogi survive his long interment. "If we compare the habits of the hybernating [sic] animals with those of the Yogis," Paul wrote, "we find that they are identically the same."

Today we know the Bengali surgeon was exactly right in terms of respiratory physiology, even if the slumber of hibernating creatures bears little resemblance to the ecstasy of accomplished yogis. Scientists have found that mammals preparing to hibernate do in fact seal off their burrows and experience the kinds of shifts in respiratory gases that Paul ascribed to yogis in their gupha. The closure of a den, notes David A. Wharton, a zoologist and author of Life at the Limits, "promotes a build-up of carbon dioxide which helps depress the metabolism of the animal," slowing its biorhythms in preparation for deep sleep.

Paul's analysis was brilliant in originality and import, his work all the more impressive given his humble status and the limited tools available to scientists of his day. He began uncovering what turned out to be one of yoga's main influences on human physiology—how the discipline can slow the metabolism.

The rapture of yogis still held secrets. But Paul had taken a bold step that began a revolution.

Investigators who wandered the Asian subcontinent in search of the miraculous had to pay close attention to the possibility of cheaters, an issue that had clearly worried the maharajah of the Punjab. India teemed with street magicians who did feats of illusion for a living—everything from charming snakes and dismembering one another to climbing ropes that disappeared into thin air. The ragtag clans had operated for centuries and had so refined their acts that Western conjurers often puzzled over the tricks and came to India to learn the secrets.

The magicians of India typically worked hard to cultivate religious associations, invoking the names of Hindu gods and saints. So, too, many poor religious figures in India—including yogis and swamis—gave in to the temptation of doing street magic as a way to make a living and often sought to pass off simple conjuring "as miraculous evidence of divine powers," according to Lee Siegel, an analyst of Indian magic.

An Indian sociologist once disguised himself as a penniless monk. His survey of hundreds of Hindu holy men found that more than 6 percent admitted to the performance of magic tricks and pseudo-yogic feats, including live burials. Interred performers would get food surreptitiously or leave the cell through a secret hole. In one case, townspeople were surprised to find an ostensibly buried holy man strolling beside a river.

In 1896, Hungary held a Millennial Exposition in Budapest to celebrate its first thousand years. The festivities were to include two holy men from India. The yogis were said to have the ability to go into deep trances, seeming to die, and then return from the dead.

Professor Aurel von Török, director of the Anthropological Museum in Budapest, was famous for his precise studies of skulls. His personal appearance echoed his love of precision and his abiding sense of caution—his beard tidy, his glasses tiny.

The professor had a difficult time making arrangements to see the holy men and taking measurements. The yogis took turns displaying themselves in a glass coffin, going into trances and switching places every week or two, always with great fanfare and prayer and incantation. The awestruck crowds and all the comings and goings conflicted with the calm required for serious investigation.

With some irritation, von Török noted in a preliminary report that true science was difficult under the circumstances. Unexpected results had clearly aroused his suspicions: the professor studied the men carefully but could find no plunge in their vital signs.

The wisdom of von Török's caution soon became apparent. A few skeptics hid themselves in the apartment with the glass coffin. Late one night, they watched in astonishment as the lid of the coffin opened and the yogi stepped out. He proceeded to enjoy a cake and a bottle of milk.

They seized the startled man.

He and the other impostor managed to escape—Houdini-like—and save their show for another city and another day.

Inquiries into the miraculous side of yoga deepened in the course of the twentieth century, as we shall see. But the analyses and exposés amounted to little compared to a potent new force that surged across the length and breadth of yoga, concealing its unbecoming aspects and bestowing on its scientific investigation new legitimacy. That, in turn, caused the sheer quantity of research to explode.

The force was Hindu nationalism. It took shape as the nation's elite, drawing on decades of rising anger over British colonial rule, worked hard to create a national identity that could unify the masses, counter notions of Western superiority, and forge the popular will necessary to oust the hated foreigners. (Around the same time, a similar effort got under way in Ireland, with similar results.)

The surge in nationalism sought to revive and modernize Hinduism as a foundation for Indian national identity, and did so across the subcontinent in countless political groups. They saw Indian antiquity as a time of cultural, religious, and social greatness. Scholars agree that the fundamental objective was to replace the myth of the white man's superiority with one of native genius.

Yoga—with its ancient roots and mystic aspirations—was seen as a potential star. But it had problems. Middle class Indians found its obsession with sex and magic to be "an embarrassing heritage," according to Geoffrey Samuel, a yoga scholar. As a practical matter, yogis had fallen so low in status that many Indians saw the unkempt drifters as symbols of all that had gone wrong with the Hindu religion.

The practice needed an extreme makeover. And it got one.

In October 1924, at the age of forty, Jagannath G. Gune established something quite new in the world—a sprawling ashram devoted to the scientific study of yoga. It was perched on a mountainous plateau south of Bombay and adjoined a hill resort where people fleeing the coast's heat would go for cool refreshment. There, amid rolling acres of lush vegetation, Gune (pronounced GU-nay) conducted what is considered the world's first major experimental investigation of yoga.

He built a laboratory, filled it with the latest instruments, hired assistants, and donned a white lab coat. He founded a quarterly journal, *Yoga Mimansa* (Sanskrit for "profound thought or meditation"), and filled it with the results of his research. True to his nationalist roots, he made sure his rambling complex had room for armies of individuals interested in yoga cures and instruction, especially young people. The ashram, he declared in a veiled reference to the Hindu drive for independence, would excel at "sending out youths that will selflessly help the building of their nation."

For nearly a half century, Gune worked with missionary zeal to direct scores of scientists who wrote hundreds of pioneering reports, helping to recast the ancient discipline as a boon for health and fitness. His toils won the admiration of Gandhi, Nehru, and many other stars of the independence movement, as well as major gurus who spread the reformulated yoga around the globe. And he did it all with a curious mix of pride and bravado, humility and innocence.

"He never wanted people to honor him," O. P. Tiwari, secretary of the ashram, told me as monsoon rains fell outside his office window. "He never wanted that people would give him credit—would say he had done a great work."

Most surprising of all, Gune came to his scientific passion not as a scientist or a physician. In terms of credentials, he was nothing close to a respected von Török, a lowly Paul, or even a student who had majored in the sciences. Nor did he have any money. What he did have—in spades— was the confidence of the independence movement.

Gune (1883–1966) had grown up north of Bombay in an area that became a hotbed of the insurrection. An orphan at fourteen, he threw himself into the nationalist struggle. He eagerly read *Kesari* (or *Lion*), a populist newspaper that urged a fallen people to boycott British goods and influence, to educate themselves, and to strive for self-rule. As a young man, he resolved to devote himself to the cause of Indian freedom through national and religious service. He vowed to remain celibate, to forgo a family, and never to serve the British. Instead of British-made textiles, he wore khadi, or homespun cloth. At one point, he roamed from village to village using the medium of Hindu music and song to spread *Kesari*'s message of independence.

His big opportunity arrived when a wealthy industrialist hired Gune for a teaching job at one of his pro-independence schools. Gune rose quickly. By 1920, his patron put him in charge of a small college. But Gune found himself out on the street in 1923 when authorities shut down the college for agitating against British rule.

His benefactor again came to the rescue. This time he gave Gune a large donation that let the jobless educator take the biggest step of his life and found the scientific ashram.

In his research, Gune made up for lost time, publishing a flurry of findings in *Yoga Mimansa*. Its language was English, signaling its wide target audience. He presented two studies in 1924, six in 1925, four in 1926, seven in 1927, and so on. Early on, he performed more than a dozen X-ray studies of yogis in various states of contortion. This surge was unique for the day.

"We cannot make even a single statement," Gune boasted, "without having scientific evidence to support it." That, of course, was a fairy tale. But it showed the depth of his enthusiasm.

The yoga taught at the ashram had been carefully repackaged. No untidiness was tolerated, nor ashes nor unkempt hair. Everything was squeaky clean—like science itself. Yoga's unsavory aspects had suddenly vanished.

Throughout his career, Gune maintained a virtual taboo on the word "Tantra"—the parent of Hatha that Hindu nationalists had come to abhor. Students heard nothing about thrills similar to "the bliss experienced in sexual orgasm," as White had put it. They got no tips about extended lovemaking, as the *Hatha Yoga Pradipika* had instructed. All that was off the public agenda. The reformulated program had to do with giving yoga a bright new face that radiated with science and hygiene, health and fitness.

Gune's investigations could be quite technical, despite his lack of formal scientific education. An early one centered on high blood pressure. The question was whether the risks of challenging poses outweighed the benefits. To study the problem, Gune had eleven students do the Headstand and the Shoulder Stand, two of yoga's most demanding poses.

Headstand, *Sirsasana*

Inversions, by definition, can unnerve. Quite suddenly, new students find their worlds upended and their hearts racing. Once beginners have achieved a measure of skill and confidence, however, they tend to find the poses strangely relaxing or, at other times, exhilarating. The conventional wisdom is that inversions reverse the effects of gravity, invigorate the circulation, and flood the vital organs and brain with nourishment, bringing about a rush of rejuvenation.

Gune and his aides found that the poses, though demanding, tended to be gentle on the heart. The traditional measure of blood pressure is how high it raises a column of mercury, and the usual daytime reading for a resting adult is around 120 millimeters. For the Headstand, Gune found that the average readings started at 125 millimeters, rose to 140 millimeters at the end of two minutes, and settled back down to 130 millimeters by the end of four minutes. That modest rise, he argued, compared favorably to how the hundred-yard dash, for instance, resulted in blood pressure soaring as high as 210 millimeters.

He wrote that the inversions still achieved the goal of "getting a richer blood supply" to undernourished parts of the body despite the "comparatively low rise" in pressure and the modest physical effort. Not that muscles were neglected. "We have ample evidence," Gune boasted, that the poses represent "an unrivalled set of exercises even for the towers of strength!"

Throughout his career, Gune showed a fondness for the zing of exaggeration. He was, after all, part showman. With the implied authority of his white lab coat, Gune worked hard to advance not only the substance of science but its style. He wanted to cultivate the idea that science had endorsed yoga—to demonstrate its approval and borrow some of its repute and progressive energy as a means of giving the discipline a new air of respectability. He desperately wanted yoga to project a new image.

But Gune also exhibited real depth. Surprisingly, given his raw political objectives and lack of formal scientific training, he repeatedly displayed a love of rigor. He even managed to disprove one of yoga's central tenets.

Yogis of his day (and ours) were happy to appear scientific by declaring that deep breathing had hidden powers of rejuvenation because it flooded the lungs and bloodstream with oxygen, refreshing body, mind, and spirit. They taught that students who did intense yoga breathing could feel the body tingle and vibrate with waves of healthful oxygen.

Not so, Gune countered after doing a pioneering set of measurements. Instead, he found that fast breathing did little to change the amount of oxygen that the bloodstream would absorb and determined that such vigorous efforts actually made their biggest impact by blowing off clouds of carbon dioxide.

"The idea that an individual absorbs larger quantities of oxygen during Pranayama is a myth," he wrote, referring to the yogic name for breathing exercises. Gune's finding might have been counterintuitive and contrary to the wisdom of the day. But it was stubbornly honest—and, as it turns out, scientifically correct.

A smart fund-raiser, Gune sent free copies of *Yoga Mimansa* to the maharajahs of India. These rich men presided over a patchwork of princely states exempt from direct British rule. Many patronized the indigenous arts and culture as part of the Hindu revival, and some had a lively interest in yoga.

His influence rose so fast and to such a degree that he quickly became a hero of the nationalist intelligentsia. By 1927—just three years after the ashram's founding—the former unemployed schoolteacher was advising Gandhi, arguably the most famous Indian since the Buddha and the most visible leader of India's fight for independence.

The issue was the pandit's health. Gandhi had had serious bouts of illness and fatigue often aggravated by his long fasts, as well as a fascination with natural cures and a disdain for Western medicine. He complained of high blood pressure. Gune recommended the calming effect of the Shoulder Stand. "In your case," he wrote in one letter, the pose "should certainly help." Gune noted that his own practice of the inversion left his blood pressure at a relaxed 120 millimeters of mercury.

Gune often promoted specific poses for particular ills and health benefits, pioneering an approach that many yogis would adopt over the decades. And he promulgated other innovations. Soon after founding the ashram, Gune, drawing on the inspiration of a martial arts mentor, established a policy of teaching yoga in classes of mass instruction. The lessons, moreover, were free.

Another novelty centered on women. At first, the ashram took in only male students. But that policy soon changed. By 1926, Gune was calling his reformulated yoga "peculiarly fitted for the females." His observation was farsighted, given the traditional male chauvinism of Hindu society and yoga's eventual popularity with women.

To say that Gune was pivotal understates the case. Even so, he remains virtually unknown in the West except among scholars. Joseph S. Alter, a medical anthropologist at the University of Pittsburgh and author of *Yoga in Modern India*, argues that he "probably had a more profound impact on the practice of modern yoga than anyone else."

Of Gune's many admirers, one of the most politically astute was the Wodeyar clan of Mysore, a city and state of southern India rich in silk and incense, coffee and sandalwood. The benevolent rajahs ruled over a realm about the size of Scotland, their ornate palace dominating the capital. Mysore was the most progressive of India's princely states, and historians say the ruling family played a skillful role in the politics of

Hindu nationalism, including the promotion of yoga as a way to build an Indian national identity.

Like Gune at his ashram, the Mysore palace sponsored a version of the ancient discipline that was far removed from the world of Tantra and eroticism. It was quite unmystical. For decades, members of the family had practiced an eclectic style that drew on Indian martial arts and wrestling as well as Western gymnastics and physical fitness techniques, including those of the British. It aimed at promoting martial culture, hardening the body, and producing feelings of pleasurable fitness.

In 1933—a decade after Gune had turned to the scientific study of yoga—the palace hired a teacher to run its yoga hall. This short man of quick temper and considerable erudition, Tirumalai Krishnamacharya, had spent his early life learning Sanskrit, Indian medicine, and other classical disciplines as part of the Hindu revival. He now developed a style that drew on the palace's gymnastic ethos.

Krishnamacharya refined postures, sequenced them with logical rigor, and combined them with deep breathing to create a fluid experience.

None of this would matter very much except that Krishnamacharya (1888–1989) produced a number of gifted students who eventually made him history's most influential figure in Hatha's modern rise. His passion and ideas about pose development led to the emergence of the Sun Salutation and eventually other flowing poses and styles, including Ashtanga and Vinyasa, Power and Viniyoga.

The Mysore palace sent Krishnamacharya on tours around India to publicize yoga, with the participants openly referring to the trips as "propaganda work." In 1934, the maharajah asked Krishnamacharya to visit Gune's famous ashram up north and study its methods. Krishnamacharya did so, traveling by train.

The following year, the palace guru adopted the theme of therapeutic benefits in his own book, *Yoga Makaranda* (Honey of Yoga), which the maharajah published. This sequel to Gune's therapeutic efforts was even more tenacious. For instance, it hailed the benefits of Utthita Parsvakonasana—a triangular pose known as the Extended Side Angle. The student bends one leg and keeps the other ramrod straight, lifting one arm over the head and bringing the other down to the floor. As a result, "pains in the abdomen, urinary infections, fevers and other diseases will be cured," the book declared with no hint of qualification, or proof.

Extended Side Angle, *Utthita Parsvakonasana*

Krishnamacharya may have been stubborn, gruff, and domineering but he trained a student who proved to be particularly important to the spread of Hatha yoga—his brother-in-law, B. K. S. Iyengar (1918–   ). The young man had been sickly all his life, and at first the yoga sessions in Mysore went poorly. Krishnamacharya soon lost interest in his new student. But Iyengar kept at it and eventually became healthy. Increasingly, like his guru, he looked to yoga for its restorative powers. He began touring India with Krishnamacharya and displaying his newfound skills, effortlessly tying his body in knots.

Iyengar, a young man of eighteen, at this point began to draw on the insights of medicine. It helped him ground his approach more deeply in the modern view of the body. His strategy was similar to what Gune and his colleagues had done—but in miniature.

The immersion began in 1936 when a surgeon by the name of V. B. Gokhale watched in astonishment as Iyengar gave a yoga demonstration and afterward helped facilitate Iyengar's relocation to the large city of Pune, which became Iyengar's home for the rest of his life. The physician became a friend, a supporter, and a knowledgeable liaison to the world of human anatomy.

Intimate knowledge of the human body—such as how its more

than two hundred bones fit together and fall into conflict—let Iyengar refine the poses. The Triangle was a good example. His method avoided subtle misalignments that could restrict movement. The beginner's pose, known formally as Trikonasana, began in a standing position as the student spread arms and legs far apart and bent the torso to one side, reaching up with one arm and down with the other. The pose was then repeated on the opposite side. The potential conflict centered on the thigh bone and a large knob at its upper end known as the greater trochanter, a spot where muscles attached. As the student bent over, the pelvis could easily strike the knobby protrusion, which stopped all downward movement. The solution was simple. It called for the student to turn the foot ninety degrees so it pointed outward. That rotated the overall leg and turned the greater trochanter backward so that the pelvis and torso were free to sweep downward. The result was a deeper thrust and a better stretch.

Triangle, *Utthita Trikonasana*

Drawing on such insights, Iyengar became a master of precision. Good alignment became his signature. He learned much about what was reasonable, what was ambitious, and what was dangerous.

Gune, who had become chairman of the Board of Physical Education in the Bombay region, saw Iyengar perform around 1945 in a public demonstration. History gives no details of the encounter between the two men—two pioneers who sought to align Hatha yoga with modern science. It does note, however, that Gune arranged for the institution where Iyengar performed to receive a financial grant.

In 1947, India won independence and the nation's powerful no longer promoted yoga as a way to build Hindu pride. Patronage ended or fell dramatically. In the mountains south of Bombay, Gune's ashram tightened its belt, uncertain of the future. In Mysore, politicians took over for the royal family.

Coincident with this plunge in domestic support, Hatha yoga went global. The exports began with a gifted student who had studied with both Gune and Krishnamacharya. Like Iyengar, the neophyte had come to yoga for reasons of ill health and had become a fervent champion of its restorative powers. Moreover, the student was a vivacious woman. She helped turn Gune's observations about yoga being "peculiarly fitted for the females" into multitudes of women devotees.

Born of a Swedish bank director and a Russian noblewoman, Eugenie Peterson (1899–2002) was a rising Indian movie star with the stage name Indra Devi when she developed a serious heart condition. She met Gune and studied at his ashram, likening it to a health spa. Devi found herself in classes with other women and distraught at her poor flexibility. A woman instructor advised patience.

The neophyte then sought to study with Krishnamacharya.

He refused.

"He said that he had no classes for women," Devi recalled.

She persisted.

Eventually the guru relented.

Devi learned well, moving to Hollywood in 1947 and teaching such celebrities as Gloria Swanson, Greta Garbo, and Marilyn Monroe. She became known as the first yoga teacher to the stars.

Devi gathered her insights into a 1953 book, *Forever Young, Forever Healthy*. It became Hatha yoga's first bestseller and the first to widely popularize the vision of ultimate health, quickly going through sixteen printings. It spoke especially to women, its tone intimate, its pages rich in fitness and beauty tips.

As yoga soared in popularity, science dug into an aspect of the old agenda that had managed to endure—veneration of the miraculous. Big claims, despite a number of exposés, had grown more prominent.

The star was Yogananda. The name of the charismatic swami meant "bliss through divine union." His book, *Autobiography of a Yogi*, told of his personal experience with yogic supermen who could fly, change the weather, read minds, walk through walls, materialize jewels, and, of no small importance to meditators in the woods, make clouds of mosquitoes suddenly disappear. It was Aladdin come true. His book, translated into dozens of languages, awed and inspired a generation of seekers. "Control over death," he declared in his writings, echoing the *Hatha Yoga Pradipika*, "comes when one can consciously direct the motion of the heart." In his *Super Advanced Course*, Yogananda gave the ostensible secret: "Yogis know how to stop heart and lung action voluntarily but keep physically alive by retaining some Cosmic Energy in their bodies."

Into this supernatural blur came something entirely new in the world of yoga exposés—a defector, a true insider who knew the field's secrets and personalities and perhaps its vulnerabilities.

Basu Kumar Bagchi (1895–1977) had grown up in Bengal, like Paul, and had enjoyed a close friendship with Yogananda. The two men went to college together, took monastic vows together, ran a school together, came to America together, preached together, and published religious tracts together. Bagchi became the second-in-command of a rising spiritual enterprise that Yogananda founded in Los Angeles. The Self-Realization Fellowship came to own many costly properties, including more than a dozen lush acres of California coastline.

The two eventually fell into bitter conflict, allegedly over Yogananda's breaking his vow of celibacy with female devotees. Bagchi gave up his monastic vows and earned a doctorate in psychology. After a stint at Harvard, he took a post at the University of Michigan and became a pioneer in deciphering brain waves for the diagnosis and treatment of disease, including epilepsy. Bagchi wrote little or nothing about yoga during this period. It was his past, not his future.

Then Yogananda died. It happened in 1952 while the famous swami was giving a talk at the Biltmore Hotel in Los Angeles. He suffered a heart attack and collapsed, his death reported on the front page of the *Los Angeles Times*. His demise at the age of fifty-nine seemed to kick the Self-Realization Fellowship into high gear. Yogananda became a departed saint. Hagiography flourished. The group released portraits of the departed yogi that fairly glowed with saintly radiance.

Bagchi now dug in. Over the course of a decade, he investigated one of the most palpable of the miracles—stopping the heart.

Bagchi recruited colleagues, won financial backing from the Rockefeller Foundation, bought the best equipment, traveled to India, visited Gune's ashram, and studied some of the world's most gifted yogis. To his delight, he eventually tracked down Krishnamacharya—the guru to the gurus who founded the main schools of modern yoga. The celebrated man had become a living testament to yogic wonders. To win converts, Krishnamacharya had taken to demonstrating what his devotees hailed as siddhis—suspending his pulse, stopping cars with his hands, lifting heavy objects with his bare teeth.

When first approached to perform the siddhis, the yogi protested. He was sixty-seven and too old. Finally, he relented. Bagchi hooked up the electrodes as the venerated yogi closed his eyes and concentrated. *Blip, blip, blip.* The recording pens flew back and forth, catching the subtle cardiac rhythms no matter how hard Krishnamacharya tried. Yes, the heartbeat was diminished. But even a quick glace at the tracing paper showed that the beat was still there, even if reduced and too faint for a stethoscope to pick up. The heart was still thumping away inside, *blip, blip, blip.*

In 1961, Bagchi and his colleagues published their findings in *Circulation*, the prestigious journal of the American Heart Association.

"It was often reported that some yogis could stop the heart," he later recalled. "Everybody including physicians thought that it was so. We discovered the truth."

Another insider joined in. He was no defector but rather a central authority in the world of yoga and one of its most respected elders.

Gune at that point was approaching his eightieth birthday, white hair spilling down his neck in curls. The cardiac studies caught his attention. After all, some of his colleagues had participated. Bagchi had stayed at the ashram much longer than anywhere else in India—more than five weeks. What the foreign scientists had come to examine and—as it turned out, to rebut—was not some trifle but a central tenet of yoga and its legacy of superhuman achievement. It put the ashram in an awkward position.

A lesser man might have denied the heart findings or disparaged them as flawed. Not Gune. Not the nationalist rebel who vowed to make no statement "without having scientific evidence to support it." So he rallied his ashram. And—to his immense credit—he did so not with reluctance or diffidence, but boldly. It was as if he, late in life, became determined to enhance the reputation of his institution and mission. Bagchi and his team had focused on the heart. Gune would take on an even bigger challenge.

Live burial was the most spectacular way that gurus and adepts had worked in public to reveal their otherworldly powers, as the Punjab yogi had demonstrated for the king.

Gune put his team into creating a samadhi pit meant to mimic the earthen dens of the miracle workers. But it was designed to minimize the chance of extraneous variables—not to mention cheating. It was dug not in a field or in sand, as yogic supermen often did, but in the foundations of a laboratory, where gas flows would be easier to monitor and eliminate. It measured six feet long, four feet wide, and four feet deep, its floor plaster, its walls brick, and everything coated in thick paint. The team installed a seal around the door to make it airtight. The precautions produced a samadhi pit that was completely sealed off from the outside world—the first of its kind. No air could enter or exhalations leave.

The ashram took volunteers from its own ranks and beyond. The most gifted turned out to be an itinerant showman of athletic build who had performed yogic feats at country fairs. He boasted of having endured live burials for up to a month. The showman, Ramandana Yogi, wore bangles on his wrists and trunks of tiger skin.

Twice in 1962 he braved the pit. The first time he managed to withstand the chamber not for anything approaching forty days and forty nights—not even for a month or a week. He went eleven hours. His second try was better. He went eighteen hours before demanding to be let out, gasping for breath.

In all, the scientists locked volunteers into the samadhi pit eleven times. Nothing like it had ever been done before. The results tore a hole in yoga's legacy of miraculous claims.

Today the ashram has a slightly dilapidated air, walls crumbling here and there amid dense foliage. But the pit is frozen in time, bright and spotless and ready for any new volunteer who might appear. It is part museum, part open challenge.

"We're still ready to do this," said Makrand Gore, a senior researcher at the ashram. He opened the pit's door while describing its past. The tidy den, well lit and brightly painted, did in fact seem ready to admit a new volunteer. A bundle of wires hung down from its ceiling, awaiting a miracle worker.

Gore's boss, T. K. Bera, a small man with a muscular presence, joined the tour. He said the ashram had looked hard for siddhis over the decades but had found no miracles—none, try as it might.

"People say yoga is black magic," Bera remarked. "But what we've found is that it gives the power to live on a reduced metabolism. That's all. It's not magic."

Popular yoga made no explicit acknowledgment of the pit demythologizing but continued to shed the old emphasis on magic and eroticism. The trend culminated with Iyengar.

His 1965 book, *Light on Yoga*, quickly became the how-to bible of Hatha yoga. Around the globe, it sold more than a million copies, confirming the field's export potential. In his preface, Iyengar poked fun at credulous people who asked if "I can drink acid, chew glass, walk through fire, make myself invisible or perform other magical acts." Instead, he described his objective as portraying yoga "in the new light of our own era."

Iyengar made no mention of Gune, Bagchi, the humiliation of his own guru, Krishnamacharya, or the coaching of his scientific tutor, Gokhale. He simply infused his book with the new sensibility.

For every posture, he noted a number of invisible health effects, often using medical terms. An example was the Locust, or Salabhasana. The student lay facedown and lifted the head, chest, and legs as high as possible. Iyengar said the pose "relieves pain in the sacral and lumbar regions" while benefiting "the bladder and the prostate gland."

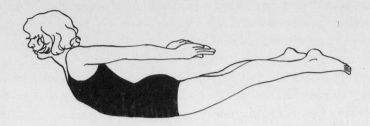

Locust, *Salabhasana*

So too he praised the Headstand. Its upending of the body "makes healthy pure blood flow through the brain cells" and "ensures a proper blood supply to the pituitary and pineal glands." Iyengar never said anything about research or clinical trials or the possibility of placebo effects. Instead, he piled on the medical terms and laid out the health benefits, giving his book a feeling of scientific authority while avoiding the messy issue of evidence. It was *light* with no explanation of its origin.

More aggressively, Iyengar claimed a wide array of cures and therapeutic benefits, again with no reference to supporting evidence other than "experiences with my pupils." His book used the word "cure" dozens of times. At the book's end, he laid out a master list of "Curative Asanas" for nearly one hundred ailments and diseases. They included arthritis, asthma, back pain, bronchitis, diabetes, dysentery, epilepsy, heart disease, insomnia, migraine headaches, polio, pneumonia, sciatica, sterility, tonsillitis, ulcers, and varicose veins.

The subliminal message was perhaps the most important of all. Nearly six hundred photographs showed Iyengar bending his supple body into all kinds of loops and curls, twists and knots. Here was an accomplished body builder whose appearance bore no hint of yoga's past. He displayed no ashes or amulets, no matted hair or beard. Ages of decay had given way to a new kind of yogi.

The agenda no longer featured sex. Even so, it still made a few appearances, often with a therapeutic spin. For instance, Iyengar put impotence on his list of curable ailments.

More important, at the very end of his book—in the final pages of a section called "Hints and Cautions," buried in a discussion of advanced practices, couched in language more evocative than explicit—he made a sudden disclosure. Even sanitized yoga, it turned out, retained a considerable measure of its old fire.

Iyengar spoke of "sexual retentive power" and suggested that the discipline could fan the smoldering embers of human sexuality into a tempestuous blaze. If the yogi gave in, he said, "dormant desires are aroused and become lethal."

It was like a doctor suddenly informing a patient that the current course of treatment had serious, previously undisclosed side effects. And it got worse. Iyengar proceeded to spell out the ultimate stakes, making his belated admission in the middle of a very large paragraph. In my edition, the disclosure comes on page 438.

Yoga, Iyengar warned, could transport the practitioner "to the crossroads of his destiny." One path led to the divine, he said, and the other to "the enjoyment of worldly pleasures."

In its fundamentals, the transformation of yoga was now complete. It had gone from the calling of supermen to the pursuit of common men—and increasingly of common women. It no longer belonged to mystic loners but to humanity. Its home was no longer India but the world. Its mode of instruction was public rather than private. To a growing degree, its practitioners no longer reveled in skulls and ashes but exercise mats and gym clothes. Enthusiasts by the millions ignored the old mysticism for the new ambitions of health and fitness. If yoga still harbored some of its old eroticism, that aspect of the discipline typically got ignored and downplayed, often to the point of invisibility.

In short, yoga had gone from an ancient obsession with transcendence of the body to a modern crusade for a new kind of physicality.

Of the ironies that come to light in a review of yoga's modernization, one of the greatest is how its health agenda—begun by Paul, seized on by Hindu nationalists, developed as an export item, marketed with bold pretense, championed globally as the ultimate life enhancer—turned out to produce a wealth of real benefits. The posturing in some respects proved to be fortuitously accurate. The evidence grows richer every few days, as suggested by PubMed's posting of new reports on yoga at the rate of more than one hundred a year.

A large body of research derives from the kinds of metabolic slowdowns that Paul began to identify a century and a half ago. For instance, scientists have found that physiological slowing from yoga can reduce stress, the heart rate, and blood pressure, helping to boost immunity and prevent diseases. In 2009, investigators at the University of Pennsylvania reported that twenty-six people who did Iyengar yoga for three months succeeded in reducing hypertension and its precursors. That is important because hypertension, or high blood pressure, is associated with an increased risk of stroke, cardiovascular disease, and kidney disease. Strange as it seems, the deathlike trance of the Punjab yogi ultimately threw light on healthy living. It turns out that several of the miracles—if false in terms of otherworldly feats—nonetheless reflect a real ability of yogis to accomplish lesser manipulations of the body that produce a range of health benefits. Take the heart. Bagchi may have shattered Yogananda's claims about full stoppage. But decades of investigations have shown that yoga can produce heart repercussions that work powerfully on behalf of cardiovascular health, a potentially vital issue of civic welfare since heart disease is *the* leading cause of death in the industrialized world. The studies range from anecdotal to rigorous. But their large number—dozens conducted everywhere from India and Japan to Europe and the United States—argue persuasively that yoga works remarkably well. It has been found to lower such cardiovascular risk factors as high blood pressure, blood sugar, cholesterol, and levels of fibrinogen, a protein involved in blood clotting. It has also reduced signs of atherosclerosis—an underlying factor in heart disease that arises when cholesterol and other fatty deposits begin to clog the arteries. Finally, it has been found to raise levels of antioxidants in the bloodstream and to lower oxidative stress, a euphemism for highly reactive species of oxygen that wreak havoc with cellular machinery.

More important, scientists have found that the lowered risk factors translate into medical benefits. Clinical studies have shown that patients who do yoga have fewer hospital visits, less need for drug therapy, and a smaller number of serious coronary events ranging from heart attacks to death. Analysts at the University of Virginia reviewed seventy of the studies and concluded in 2005 that yoga shows promise as a "safe and cost-effective intervention" for improving cardiovascular health.

A different field of research shows that yoga can counteract the forces of aging—another area rife with miraculous claims, as when the *Hatha Yoga Pradipika* spoke of eliminating wrinkles and gray hair.

Consider a 2011 study. It looked at elderly women who took up yoga and found their balance much improved. That's significant. For seniors, falls are *the* leading cause of death by injury.

The spine is another target. Yoga has long claimed that all the bending and stretching will make the backbone youthful. Science has examined such declarations and found that yoga can, in fact, counteract the deterioration of the disks that lie between the vertebrae.

The watery cushions act as pivot points and shock absorbers so that the vertebrae can move smoothly, letting the body go through its regular bending and rotating. The disks of adults have no blood supply of their own but instead rely on nearby vessels to nourish them. With normal aging, the already limited supply of blood diminishes still further and the disks gradually dry out and become thinner. As a result, the trunk shortens and a person shrinks in size. The thinning of the disks can result in a number of nerve conditions and severe pain.

In 2011, the idea that yoga can slow such deterioration received support when physicians in Taiwan reported on a study of thirty-six people. Half had taught Hatha yoga for at least a decade, and the other half were judged to have exhibited good health. The two groups showed no statistical difference by age or sex. The physicians then scanned all the spines and carefully inspected the disks for signs of damage. The results, the team wrote, showed that the yoga teachers had "significantly less" degenerative disease than the control group.

Why? The physicians suggested that spinal flexing may have caused more nutrients to diffuse into the disks. Another possibility, they wrote, was that the repeated tension and compression of the disks stimulated the production of growth factors that limited aging.

The frontiers of biomedical science turn out to hold many clues to prospective health benefits. The new understandings reveal potential—if unproven—rewards for practitioners even if the word "yoga" never appears in the text or title of a scientific paper.

One surprise centers on the vagus, often portrayed as the most important nerve in the body. It travels from the brainstem to the torso, where it radiates out to the lungs, heart, stomach, liver, spleen, colon, and other parts of the abdomen. The word *vagus* shares etymological roots with "vagrant" and "vagabond," denoting how it wanders through the body.

The nerve's action is central to the regulation and slowing of the human heartbeat, and thus has played important roles in ostensible miracles going back to the days of the Punjab yogi. But the new research focuses on what turns out to be an even more fundamental talent of the nerve—the regulation of the immune system, in theory offering protection against a number of serious illnesses.

The body's immune response is typically portrayed as white blood cells battling foreign invaders, and the immune and nervous systems as distinct entities—like oil and water, never mixing. The itinerant nerve would thus seem to have nothing whatsoever to do with the body's defense mechanisms.

Kevin J. Tracey found otherwise. In 2002, the immunologist at the North Shore University Hospital on Long Island, New York, reported that the vagus wields remarkable control over the body's immune system, playing major roles, for instance, in fighting inflammation.

That may sound unimportant. But a number of deadly conditions arise from the body's overreaction to infection or its threat. For instance, the whole body can swing into an inflammatory state known as sepsis, a quiet killer that in the United States takes more than two hundred thousand lives each year. Other disorders include lupus (an autoimmune disease), pancreatitis (a chronic inflammation of the pancreas), and rheumatoid arthritis (a chronic inflammation of the joints). Scientists are working hard on anti-inflammatory therapies.

Tracey initially focused on drugs meant to excite the vagus. But the more he learned of yoga and other Eastern disciplines, the more interested he became in their potential as natural agents to fight inflammation and its debilitating effects. In 2006, he discussed the topic at a conference held under the auspices of the Dalai Lama, the Tibetan spiritual leader who has long shown an interest in Western science.

Tracey's ideas won support in 2011 when Indian scientists under the leadership of Shirley Telles—one of yoga's most prolific investigators—reported that doing intensive practice for a week can ease trauma from rheumatoid arthritis, the painful disorder of the joints. It afflicts millions of people. The study looked at sixty-four patients, ranging in age from twenty to seventy. The yoga included flexing poses and slow breathing, which stimulates the vagus. Measurements at the beginning and end of the week showed drops in rheumatoid factor—an indicator in the bloodstream of the disease—as well as improvements in the ability of practitioners to get out of bed, dress, walk, eat, and grip objects.

Investigators of the invisible are finding even deeper allures. They include an ultimate expression of good health—longevity.

Few topics in yoga have produced more fog. The mythology goes back at least as far as Marco Polo, who first visited India around 1288 and reported that yogis could live for as long as two centuries. Today, yogis and yoga teachers routinely hail the practice as greatly prolonging life—though no study that I know of has examined that claim. What makes headlines are anecdotes. For instance, many authors point to the longevity of Krishnamacharya, who became a centenarian. So too his student Indra Devi, author of *Forever Young*, drew attention by living to be one hundred and two. Few yoga enthusiasts mention that pudgy Yogananda died of a heart attack before he was sixty.

Despite the wishful thinking, a recent discovery suggests that yoga can indeed slow the biological clock. The finding centers on a longstanding riddle—why cells age, die, or in some cases defy the natural order of things to remain young. The answer involves the microscopic whorls of DNA that lie at the tips of the chromosomes, the central repositories of genetic information in the cells. Scientists have found that these DNA tips, known as telomeres, get shorter each time a cell divides and thus serve as a kind of internal clock that determines the cell's allotted time in life. They have also discovered the secrets of telomere growth and youthfulness. The finding was considered so important that it won the 2009 Nobel Prize in Physiology or Medicine. To scientists, the story of the telomere suggested a more accurate way of measuring biological age than simply marking the passage of the years.

As often happens in science, the discovery brought into sharp focus yet another question—why do the telomeres of some individuals hold up much better than others? In some cases, an eighty-year-old could have the long, youthful telomeres of a thirty-year-old. Why the variation?

It turned out that a number of everyday conditions eroded the telomeres—a main one being chronic psychological stress. (Other factors include unhealthy diets and infections.) Happily, science also found that reducing stress could slow the biological clock. The slowdown was found to work even with subjects well into their middle and later years. Perhaps most intriguingly, given humanity's long search for a fountain of youth, a few tentative studies suggested that short telomeres could be coaxed into growing long again, in effect turning back the biological clock.

Enter yoga. Science over the decades has repeatedly shown yoga's talent for undoing physical and mental stress, as we will discuss in chapter 3. Thus yoga, despite its checkered history on longevity claims, appears to be custom made for slowing the biological clock.

Dean Ornish led the appraisal. A Harvard-trained physician known for his popular books, Ornish was a longtime devotee of yoga, having begun his practice in the 1970s. Over the years, he developed and marketed a health plan that championed a combination of yoga, low-fat diets, whole foods, and relaxation techniques. Studies of his method became part of the evidence for yoga's cardiovascular benefits. Now he turned his attention to the telomeres, in particular to a measure of their maintenance and building known as telomerase—an enzyme that adds DNA at the chromosome tips. He did so with colleagues from the University of California at San Francisco, including Elizabeth Blackburn, who was soon to share the Nobel Prize for her telomerase findings.

The team looked at twenty-four men who took up the Ornish program. They ranged in age from fifty to eighty and did yoga for an hour a day, six days a week. The scientists assessed telomerase levels and other physical and psychological measures before the men began their overhaul and did so again at the conclusion of the three-month program. The results were unambiguous. The scientists found declines in cholesterol, blood pressure, and such indicators of emotional distress at disturbing thoughts. More important, they discovered that levels of telomerase shot up 30 percent.

The team reported its findings in late 2008, proclaiming them a first. The eleven scientists said the findings had implications for cellular longevity, tissue renewal, disease prevention, and "increases in life span"—a holy grail of modern science.

The Ornish inquiry was only a beginning, of course. Other investigators would have to zero in on yoga practitioners and do larger and more elaborate studies. But it was a start.

• • •

As science over the decades succeeded in promoting health over the miraculous, some yogis nonetheless managed to cling tenaciously to the past and show a recurring fondness for discredited myths. Major gurus gave up wild declarations. But other authorities were often quick to embrace lesser miracles and trendy fictions.

The Complete Idiot's Guide to Yoga proclaimed that yogis can "live to be well over 100 years" and can "stop their own hearts (then start them again, of course)."

So, too, Georg Feuerstein, a star of yoga scholarship, concluded that science "lent credibility to many" of yoga's more astonishing claims, making them "appear far less outlandish than they once seemed."

Feuerstein's essay on the topic appeared in his book *Sacred Paths* and was entitled "Science Studies Yoga." But his enthusiastic tour of the field managed to say nothing about the decades of demystification. Nor did it mention the specific exposés of Paul and von Török, Gune and Bagchi, Gore and Bera. His message was all about supermen.

As it turned out, claims of the fabulous went beyond the rarified world of the advanced yogis. They expanded in time to include everyday practitioners as well.

II

FIT PERFECTION

once took a yoga class where the male teacher held up an illustration of a muscular man and proceeded to ridicule the build as childish. His message was loud and clear—yoga produces a better physique and is more advanced than other varieties of body development. The class nodded in agreement. We were all in the same boat headed for the same destination—a place of lithe contours and sculpted abs, a land where physical fitness can reach astonishing new heights.

Of course many yoga teachers honor the field's spiritual roots with attitudes of humility. Some acknowledge that yoga has its own strengths and limitations. Even so, a number of gurus and teachers have put forth extraordinary lists of particulars to explain why yoga constitutes the ultimate form of exercise.

Consider Bikram Choudhury, the founder of hot yoga, a man famous in the yoga community for his collections of Rolexes and Rolls-Royces. His brand is so popular and uniform that some call it McYoga. He demands that every studio do exactly the same sequence of twenty-six poses and two breathing exercises.

Choudhury grew up in impoverished Calcutta but struck it rich in the United States, opening many hundreds of yoga centers. Uniquely, the exercise room of a Bikram studio is heated to a sweltering 105 degrees Fahrenheit (not unlike Calcutta on a summer day). Gleefully, Choudhury calls it his torture chamber—and indeed, beginners who enter the mirrored halls often experience spells of dizziness and nausea. Some pass out. The underlying theory seems to be that heat loosens the joints, muscles, and tendons and helps intermediate students push themselves hard, giving them a gratifying sense of progress.

"So many Americans," Choudhury scolds in his book *Bikram Yoga*, "ruin their bodies by blindly running around 'exercising' and playing sports. I tell my students, 'No barbells, no dumbbells, no racket.' Games are okay for children, for recreation and to teach them sportsmanship. But after that, you must give up trying to put a little round ball in a hole all the time."

In great detail, Choudhury explains why his yoga is superior to every other type of physical workout and why it deserves your attention and—perhaps most important—your money. Remarkably, he even rejects all other styles of yoga. A standard estimate for the number of people in the United States who do yoga is twenty million, and Choudhury happily cites that number as representing a world of misguided souls.

"Bogus yoga" is what he calls their practice. He ridicules other approaches as watered down to accommodate American weakness and inflexibility. Among the competition, he scoffs at Kundalini, Ashtanga, and Vinyasa ("which never existed in India"), as well as Iyengar ("he uses so many props in his method that he's called 'The Furniture Yogi' in India"). The newer yoga brands, he added, are even more ridiculous. "You've got Easy Yoga, Sit-at-Your-Desk Yoga, Yoga for Beginners, Yoga for Dummies, Yoga for Pets, and Babaar [sic] Yoga. It's all Mickey Mouse Yoga to me."

The false prophets, he charges, shirk their responsibilities to ancient tradition and cheat students out of "the perfect life," keeping them from the rewards of "optimum health and maximum function." In contrast, he portrays his own style in cartoonish superlatives: "You'll become a superman or a superwoman!"

Is he right? Is there more substance to hot yoga than Bikram's boastfulness would imply? And what of the other styles? Are there objective measures that can establish the benefits and compare them to regular exercise and sports? In short, is it possible to find out what is real and what is not?

While the scientific investigations of yoga over the decades have tended to be sketchy and idiosyncratic, the subject of physical fitness is one area that has received a fair amount of scrutiny. The reason has to do with the prickly nature of intellectual turf.

The academic world has a number of research fields that lavish attention on questions of fitness. The disciplines include biomechanics, kinesiology, exercise physiology, nutrition, physical therapy, and sports medicine, among others. Today, sports scientists draw on a wealth of instrumentation and software to conduct careful studies of exercise and athletics. Whole businesses do nothing but sell the equipment. Major universities have whole departments that do nothing but conduct fitness studies and publish the results in dozens of specialty journals, including the *Journal of Exercise Physiology* and *The American Journal of Sports Medicine*. The field's textbooks tend to be gargantuan in size and extraordinarily detailed in content. The professional societies of the field include the American College of Sports Medicine—the world's largest organization of scientists devoted to the study of athletics. Its standing is such that governments around the globe routinely adopt its guidelines for physical activity in campaigns meant to promote public health.

Yoga's fitness claims fall squarely into this whirl. A philosopher would say they fall within an existing paradigm. The situation is very different from the case with yogic declarations about, say, body currents and spiritual renewal.

As a result, a relatively large number of scientists (a growing number of them yogis) have applied the instruments and the techniques of the academic sports establishment to the study of yoga's fitness claims. The results, as we shall see, raise significant doubts about some of modern yoga's most prominent declarations.

A complicating factor is that yoga, taken as a singular activity, represents an oversimplification rooted in the discipline's timeless image. A better word would be "yogas," denoting the evolution of many styles over the centuries—with new ones appearing all the time. Three phases stand out. First was the original Hatha, which debuted as a forceful branch of Tantra. Then, as we saw in the previous chapter, the yoga innovators of the early twentieth century produced a sanitized Hatha. Today, the newest styles represent another step in yoga's development, their moves more vigorous than the old. It turns out that modern yoga, by accident or design, has lost much of its contemplative nature and adopted some of the sweatiness of contemporary exercise.

Gune taught a style of yoga that epitomized the slow, tranquil approach. His emphasis was on holding poses for long periods of time and learning how to relax even amid extreme states of bending, flexing, and upending. It was a point he drove home with his measurements of how challenging inversions were gentle on the heart. By contrast, the newer styles tend to be hyperkinetic, some done to the beat of rock music. The objective is to get the heart pounding and the body exhausted. That makes them more aerobic ("requiring air")—in other words, more focused on reinvigorating the blood. In contrast to Gune's style of yoga, the new goal is to maximize rather than minimize the energetic costs.

Brands that focus on aerobics include YogaFit and Power Yoga. To a lesser extent, the vigor extends to older styles such as Ashtanga and Vinyasa. And then there is Bikram. "My classes are so hard," Choudhury boasts, "you use your heart more than if you run a marathon."

Fortunately, that kind of pronouncement is open to investigation.

The analytic lens of the sports establishment began to form in the nineteenth century as health authorities struggled to identify universal factors that determine the origins of human fitness. The question was seen as urgent. Around the globe, waves of people were leaving farms and giving up agrarian lifestyles that had kept them physically active from dawn to dusk. Medical experts agreed that the new sedentary lifestyles of the cities were often unhealthy but could achieve no consensus on what forms of exercise to recommend—even as entrepreneurs and hucksters got rich promoting their own methods. It was an age of dumbbells and medicine balls, of weighted clubs and chest expanders, of gimmicks and gadgets. The scientific goal was to develop objective standards that would let investigators cut through the competing claims and document what was truly beneficial. The resulting programs of exercise were seen as important to help city dwellers improve their health, avoid fatigue, and better enjoy their lives and leisure time.

By 1900, investigators had identified a factor that they called vital capacity. It measured a person's ability to breathe deeply—seemingly a good measure of fitness because breathing is considered a foundation of the metabolism and, in earlier days, was viewed as an expression of the human spirit and soul. Science saw deep breathing as similar to blowing on a fire—in theory it fanned the body's metabolic flames.

Seeking precision, scientists defined vital capacity as the maximum volume of air that an individual could exhale after a deep inhalation. A sedentary life was found to reduce vital capacity, and an active life to increase it. Scientists quickly developed a refinement known as the vital index, which sought to eliminate differences due to age, size, sex, and other individual factors. It consisted of the ratio of vital capacity to weight. Early in the twentieth century, athletes aspired to a high vital index as an indication of competitive excellence.

Gune became an enthusiastic fan of the vital index and cited yoga's impact on the physiologic measure as evidence of the discipline's power to raise human vitality. Viewed narrowly, his claims were exactly right. Pranayama gave the lungs, the chest, and the abdominal muscles a comprehensive workout and improved the flexibility of the rib cage. The natural result was an aptitude for deep breathing. The big question was whether the pulmonary skills translated into heightened fitness.

Gune had no doubts. In his estimation, yoga, with its proven ability to expand the lungs, outshone all other sports and systems of exercise. And he said so bluntly. Shortly after starting his ashram, he declared that the discipline excels at "increasing the vital index" and improving all aspects of life. Yoga, Gune insisted, let students attain the "physiological perfection of the human body"—not improvement or development but *perfection.* "There can be no other system more suitable."

Unfortunately, just as the guru was seizing on the vital index as evidence of yoga's superiority, scientists in Europe and the United States were abandoning the measure as deceptive and potentially meaningless. For instance, they noted that the vital index of a growing child usually fell steadily between the ages of, say, ten and twenty, since body weight during those years increases faster than lung size. Yet common sense suggested that those same years saw great rises in athletic prowess.

So the question arose with new urgency: What did, in fact, define the human capacity for physical vigor and, if such a factor existed, could science find a way to measure its development?

In the 1920s, as Gune was beginning his program of experimentation, some of the world's best minds took up that question. A star was Archibald V. Hill, an English physiologist who won the Nobel Prize in 1922 for showing how muscles use energy.

Thirty-seven at the time of the award, Hill wore a proper British mustache and was married to Margaret Neville Keynes, the sister of the economist John Maynard Keynes and a social worker who had written extensively on child labor. The couple had two boys and two girls. Hill, as it turned out, was just starting a long, productive career. After his Nobel work, he turned to the related question of how muscles get their oxygen. It was the flip side of the energy coin—focusing on origins rather than ends. His agenda was quite sensible for an ambitious scientist curious about the fundamentals of biology.

Hill brought to his research an abiding personal interest in sports and fitness. As a young man, he had run competitively, covering two miles in a little more than ten minutes—a fast pace for the day. As an adult, Hill often ran a mile before breakfast. For his studies of oxygen, Hill and colleagues designed experiments meant to reveal the exact dimensions of its invisible uptake. His main venue was a grassy track. His runners strapped to their backs bags into which they would breathe at set intervals. Later, analysis of the contents revealed the quantity of oxygen consumed.

Careful measurements showed that the runners—once achieving a certain intensity of effort—could increase their oxygen uptake no more. The situation held steady no matter how much they sped up their pace or how hard they pushed themselves. It was a hidden barrier. Like a bellows blowing air, the heart and lungs turned out to work beautifully at fanning the body's inner fire but had intrinsic limits that no level of effort could overcome.

In pioneering reports of 1923 and 1924, Hill and his colleagues coined the term "maximal oxygen uptake," defining it as the peak consumption of oxygen during exercise that got incrementally harder. It soon became the gold standard of physical fitness and exercise physiology—the single most important factor in determining what made for athletic excellence. The vital index, meanwhile, was cast onto the scrap heap of history.

What determined the maximum uptake? Amazingly, peak oxygenation of the body was found to have little or nothing to do with lung size, lung elasticity, the depth of breathing, eating habits, vitamins, the amount of sleep, good posture, body weight, or whether an individual possessed an unusually potent form of hemoglobin or some other energizing factor in the bloodstream. No. The scientists concluded that it rested on one main factor—the size of an individual's heart and its ability to send blood rushing through the lungs and blood vessels to the muscles. In short, the secret of athletes who drove themselves to heights of physical performance centered on a big heart.

A central myth of Hatha yoga—one Gune had identified—held that deep breathing increased the blood's oxygenation despite the relative stillness of the body and the modest use of the muscles during yogic practice. Hill ignored that misunderstanding. His discovery centered on the *quantity* of blood oxygenation rather than mythic attributes of *quality*. It bespoke huge volumes of rushing blood. Peak oxygen consumption was typically expressed in liters of oxygen—with top athletes each minute drawing in six, seven, or even eight liters—in other words, up to two gallons. *Two gallons.* It was a flood compared to a phantom trickle. With great elegance, Hill and his colleagues overturned the misconceptions of the vital index to show that the central element of peak oxygenation rested on the workings of the heart rather than the lungs.

Today in sports medicine and exercise physiology, peak oxygen consumption is known by the ubiquitous acronym VO_2 max. In the argot of science, the V stands for volume, the O_2 for oxygen in its usual chemical notation, and "max" for maximum. VO_2 max is accepted around the globe as the best single measure of cardiovascular fitness and aerobic power.

In the early days, the question was whether coaches and individuals could raise the maximum uptake so as to increase athletic performance. The answer emerged quickly: very much so. Regular aerobic training turned out to increase the size of the heart, most especially its left ventricle—the heart's largest chamber, which pumps oxygenated blood into the arteries and body. A bigger left ventricle sent out more blood per beat and more oxygen to the tissues and muscles. Scientists sought to measure the rise. It turned out that the cardiac output of elite athletes was about twice that of untrained individuals.

The benefits extended to most anyone who took up vigorous exercise. In time, scientists found that three months of endurance training could raise VO_2 max between 15 and 30 percent. Two years raised it as much as 50 percent.

The new perspective was a breakthrough. At last, after many decades of mistakes and misapprehensions, scientists had uncovered what seemed like a dependable guide to human fitness.

The topic was long obscure. Then Kenneth H. Cooper came along. A track star in his native Oklahoma, the physician worked for the Air Force and early in his career devised a simple test that provided a good estimate of an individual's VO_2 max. The test measured how far a person could run in twelve minutes. Cooper's rule of thumb let the Air Force quickly assess the fitness of new recruits. Eager to popularize his insights, he invented a new word, "aerobics," and in 1968 authored a book by the same name. It drew on his years of research to show what kinds of exercise produced the best cardiovascular workout. Cooper found that such muscular activities as calisthenics and weight lifting were the least effective. Participant sports like golf and tennis came in second. And the big winners? Challenging sports like running, swimming, and cycling, as well as vigorous participant sports such as handball, squash, and basketball. His analyses caught on rapidly and helped get millions of people off their chairs and into the streets. Starting in the 1970s, jogging became fashionable.

The surge of activity resulted in a number of scientific inquiries that examined what aerobic exercise could do not only for athletics but health. The results were dramatic. Perhaps most important, the studies showed that aerobic exercise lowered an individual's risk of heart attack and heart disease—*the* leading cause of death in the developed world. It also reduced the prevalence of diabetes, stroke, obesity, depression, dementia, osteoporosis, hypertension, gallstones, diverticulitis, and a dozen forms of cancer. Finally, it helped patients cope with all kinds of chronic health problems. Frank Hu, an epidemiologist at the Harvard School of Public Health, praised the benefits as exceptional. For general health, he called vigorous exercise "the single thing that comes close to a magic bullet."

Why did it do so much good? Scientists found that forceful exercise improved the performance of virtually every tissue in the human body. For instance, it produced new capillaries in skeletal muscles, the heart, and the brain, increasing the flow of nutrients and the removal of toxins. Scientists also discovered that it raised the number of circulating red blood cells, improving the transport of oxygen. Still another repercussion centered on blood vessels. It caused their walls to produce nitric oxide, a relaxant that increases blood flow.

The wide health benefits prompted medical groups to call for regular exercise and public institutions to set recommended levels. The American College of Sports Medicine said healthy adults should engage weekly in at least three vigorous exercise sessions, each twenty to sixty minutes long. The American Heart Association called for at least five sessions. Many other groups, including the President's Council on Physical Fitness, made similar recommendations. The push was global. In Geneva, the World Health Organization said regular aerobic exercise held out the promise of "reducing cardiovascular diseases and overall mortality," the rate at which people die.

In short, vigorous exercise for health maintenance and enhancement became a modern credo. The message was etched in stone. Experts might quibble over the amounts. But they agreed on the principle and did whatever they could to promote its public acceptance.

It took decades for scientists to consider how yoga measured up. Part of the problem was the relatively small size of the yoga community and its limited ability to win scientific attention. Another was the difficulty of monitoring the aerobic status of practicing yogis. It was easy for investigators to study how yoga could increase an individual's flexibility and muscular strength—fair measures of fitness. But gauging the flow of invisible gases was a different story. That kind of information was hard enough to get with athletes working out on treadmills. The investigators had to fit their subjects with clumsy face masks and tubes that delivered the gaseous flows to measuring devices. But with yoga—given its range of motions and its series of rather profound rearrangements of the human body—the challenge was far greater. Even so, a number of scientific teams made headway over the years.

Cooper, the VO_2 max popularizer, did no direct investigations of yoga but carefully examined several activities that were similar, including isometrics and calisthenics. His verdict? They did little or nothing to strengthen the heart and raise oxygen consumption.

"Is your chest heaving?" he asked of the person doing the muscular tensing of isometrics. "Is your heart pounding? Is the blood racing around your system trying to deliver more and more oxygen? Nonsense. None of these beneficial things is going on, nothing that anyone can measure, anyway. We tried it and failed."

Yoga's social rise in the 1970s and 1980s led scientists to start assessing how it measured up against aerobic sports. As fate would have it, one of the first investigations was also one of the best. It was done by scientists at the Duke University Medical Center, in Durham, North Carolina, a top institution for biomedical research. The team studied nearly one hundred older adults—forty-eight men and forty-nine women. A third did Hatha yoga, a third exercised on stationary bicycles, and a third did nothing out of the ordinary.

The team's use of experimental controls set the study apart from what specialists consider an underworld of shoddy research. Controls let scientists zero in on a single variable and avoid subtle misunderstandings. They try to eliminate the complexities of nature and human interaction to ensure that any observed changes are the result of the examined factor rather than some extraneous influence. With the Duke study, for instance, the experimental controls let the scientists make sure that the process of simply gathering the subjects to the site of the investigation played no role in the results. What if some walked there? What if some bicycled? What if some ran? Would that affect the fitness measurements? The changes observed in a control group could alert scientists to the existence of an unintended influence and help them eliminate it from their findings. The big challenge for a scientist designing a study with human subjects is to make the experiences of the experimental and control groups as similar as possible—with the exception of the issue under examination. Without such precautions, researchers have no way of knowing whether the changes observed in an experiment would have happened anyway. The practical difficulty of such precautions is their added expense. The recruitment of more subjects—and their subdivision into different kinds of activities—can result in the need for more money, more personnel, more data analysis, and more administrative burdens. But the scientific benefits are usually seen as worth the costs.

In the Duke study, the hundred or so subjects, including the control group, did their designated activities for a total of four months. To get around the measurement dilemma, the team made no readings during the months of assigned activities and instead opted for detailed assessments before and after the training.

The results, published in 1989, were unambiguous. The aerobics group improved its VO_2 max significantly, raising peak oxygen consumption by 12 percent. But the yogis showed no increase whatsoever and in fact registered a bit of a decline, though it was judged to be statistically insignificant.

A surprise also emerged. The scientists were intrigued to discover that the yogis, despite their poor showing in terms of aerobic conditioning, nonetheless felt better about themselves. The subjects reported enhanced sleep, energy, health, endurance, and flexibility. They described how they experienced a wide range of social benefits, including better sex lives, social lives, and family relationships. Psychologically, the scientists said, the yogis reported a number of improvements. They had better moods, self-confidence, and life satisfaction. With few exceptions, they said they looked better.

The Duke findings hinted at a fascinating split. It was one thing to *do* good for the hidden intricacies of human physiology and quite another for an individual to *feel* good about themselves. It was the difference between improved fitness and outlook. The subjects who did yoga *felt* they had received a wealth of benefits even though the Duke scientists found no indication whatsoever of aerobic gains. Their discussion of the research findings hinted at their fascination. The improvements in attitude, the scientists said, "are worth noting."

The Duke team—unknowingly—had stumbled on one of yoga's secrets. The next chapter will explore the science of how the discipline lifts the human spirit.

Yoga fared slightly better in subsequent studies of aerobic conditioning. One reason was a subtle change in the discipline that put growing emphasis on energetic poses and styles. The new forms downplayed stationary postures for ones that required a much greater level of movement

Sun Salutation, *Surya Namaskar*

and physical activity, creating a more athletic experience and increasing the aerobic challenge.

To a surprising degree, the new vigor centered on a single activity— Surya Namaskar, Sanskrit for "salutation to the sun." Today it is one of yoga's most popular poses. The student, rather than remaining motionless in a fixed posture, moves through a fluid series of up to a dozen interconnected poses that go from standing to bending to lying prone to standing back up and to stretching backward. If done rapidly—and repeatedly—the sequence can leave the heart pounding and the lungs gasping for air. It therefore has elements of a cardiovascular workout.

The Sun Salutation and its relatives are, by nature, quite malleable. They can be sped up or slowed down to suit individual preferences. In their adaptability, they are quite different from yoga's static postures. The situation is similar to what we experience in terms of gait. When standing motionless, we are, by definition, stationary. But once in motion, we can move forward in a number of ways: walking, jogging, running, or racing ahead as fast as we can. It depends on what we want to do.

The Sun Salutation appears to be fairly recent in origin. The *Encyclopedia of Traditional Asanas*—published in India by the Lonavla Yoga Institute, founded near Gune's ashram by one of his students—draws on nearly two hundred books and unpublished manuscripts to describe many centuries of pose development but says nothing about the Sun Salutation. So, too, the asana makes no appearance in the how-to guides of Gune (1931), Sivananda (1939), and other early teachers.

The pose most likely arose in the early twentieth century as the Mysore palace and Krishnamacharya mixed traditions of British gymnastics and native wrestling. Whatever its exact origins, the Sun Salutation debuted as an important new feature of Hatha yoga in the 1930s, spreading slowly through India and the world. The idea behind the pose and kindred postures was what Krishnamacharya called Vinyasa (*vi* denotes "in a special way" and *nyasa* "to place"). It stood for the flowing movements that he developed to join the individual poses into a new kind of graceful activity. The result was a kind of yoga ballet.

In the West, students of yoga learned about the pose in a number of ways. Krishnamacharya's student Sri K. Pattabhi Jois played an important role in popularizing the series of movements and the Vinyasa system, calling it Ashtanga (or eight-limb) yoga, after the sutras of Patanjali and their eight rules. Starting in the late 1960s, Westerners began traveling to Mysore to study yoga with Jois. Slowly Ashtanga grew in popularity, especially among the physically ambitious in the West who were seeking yoga's most athletic expressions. The aggressive style required skill and power, and could leave a student bathed in sweat.

Science looked into Ashtanga as the style gained in popularity and found that, compared to traditional yoga, it posed a greater challenge to the heart. One study examined sixteen volunteers. The human heart beats about seventy times per minute. On average, the hearts of the yogis quickened to ninety-five beats while doing Ashtanga, compared to eighty beats during conventional Hatha. The Ashtanga factor represented a rise of roughly 20 percent.

The more difficult question was whether the increased thumping of the heart that resulted from faster poses and faster styles translated into measurable improvements in cardiovascular fitness. That soon became *the* question.

Ezra A. Amsterdam was hitting new highs in his career when yoga caught his eye. A senior cardiologist at the medical school of the University of California at Davis, he had devoted himself to the study, practice, and teaching of ways to prevent heart disease, the nation's number one killer. He was prolific. His résumé boasted hundreds of articles. Recently, he had even founded a journal—*Preventive Cardiology*—the first of its kind, published by John Wiley & Sons, a respected firm. Amsterdam's own studies ranged from investigations of diet and exercise to drugs and therapy as ways to promote a healthy heart and fend off cardiovascular disease. He lived in sunny California and practiced what he preached, maintaining a trim figure into his sixties.

Yoga as a field of scientific inquiry seemed wide open to Amsterdam, his interest stimulated by what he saw as "the lack of objective study." Growing numbers of people were practicing yoga, and doing so in new, innovative ways that were often quite vigorous. A fresh look at the relationship between yoga and aerobics seemed to beckon.

Another factor was his daughter. Dina suffered from an eating disorder and was twenty-five pounds overweight when she starting doing yoga. It worked like a charm. The discipline helped her shed weight, feel good about herself, and get in excellent shape. She studied with Rodney Yee, a yoga star, taking his advanced teacher training course. Dina became a devotee of the discipline and proceeded to teach classes in the San Francisco area. She had a big smile and a reputation for rigor, sensitivity, and infectious enthusiasm. Dina—a graduate student at Stanford University—also had a deep interest in the science of yoga and wellness. She had "many lively discussions" with her father, she recalled, and was delighted when he decided to do a yoga study.

Amsterdam was determined to give yoga a serious look. Was the discipline all that it was cracked up to be? Was it, in fact, all that Dina needed to stay fit, to maintain a healthy heart? Could any practitioner reap the benefits? If so, yoga might join the elite club of rigorous sports and activities that public health authorities had singled out as highly beneficial—especially in preventing heart disease and the kind of cardiovascular illnesses that Amsterdam knew only too well.

He worked with a team of specialists from the University of California at Davis, a good school in a highly respected system. Except for him, the three researchers came from the department of exercise science, anchoring the study in a solid analytic tradition. In terms of capabilities and intellectual depth, the team appeared to be quite strong.

But the investigation, begun on an auspicious note, soon encountered a number of difficulties. The biggest was the lack of major financial support for the study, which forced the scientists to limit its size and design. They lined up just ten volunteers—one man and nine women. Compared to the Duke study, that was one-tenth the number of subjects. Moreover, their examination had no control group. The low numbers and the absence of controls increased the possibility that any observed changes might result from random variability rather than yoga. A final limitation was that the students were required to do a minimum of just two workouts a week for two months—a fairly short time in which to see the physiological effects of yoga. By contrast, the Duke study had proceeded twice as long.

Even so, the yoga session itself was fairly intense, lasting nearly an hour and a half. It included ten minutes of breathing exercises (pranayamas), fifteen minutes of warm-up exercises, fifty minutes of yoga postures (asanas), and ten minutes of relaxation in the Corpse pose (Savasana). A centerpiece was the Sun Salutation. The students did two or three cycles of its fluid movements, stretching and bending back and forth. In addition, the workout featured a number of other vigorous moves that went beyond yoga's tradition of stationary poses. They included lunging forward on the legs and bobbing up and down in what the investigators called the Frog.

Different schools of yoga mean different things when they talk about the posture. The energetic pose adopted by the Davis team was a newcomer to yoga, its origins unclear. No classic text mentions its repetitious movements. It starts with students squatting down, putting their hands on the floor, and then straightening out their legs. While raising their bottoms high in the air, they keep their heads as close as possible to their knees. The movement ends with the students lowering themselves back down into the squat. Modern texts that describe that style of Frog recommend doing it anywhere from fifteen to more than one hundred times, its rhythms growing increasingly fluid and fast as the student warms up.

The ten volunteers in the Davis study had led fairly sedentary lives. A condition for participation in the study was that they had engaged in no regular physical activity—including yoga—for the previous half year. Moreover, the researchers had the students refrain from all other forms of exercise. As with the Duke study, the researchers got around the measurement problem by performing the physiological assessments before and after the yoga training.

Having gathered and analyzed the data, the Davis team got ready to present its findings to the world. That meant finding a reputable journal.

Not all public representations of science are created equal. Journals range from bad to great. A minimum requirement for a good journal is that it conducts a process known as peer review—that is, it maintains panels of scientists working in the field who review any proposed article. They exercise what amounts to quality control, making sure a submission hangs together and, if weak, gets rejected or revised to address the inadequacies. Some of the world's best journals are published by professional associations and have long histories. *Science*, for instance, was founded in 1880 with the financial support of Thomas Edison and is now published by the American Association for the Advancement of Science, a large professional group headquartered in Washington. The best journals—the ones most widely accepted and admired within the scientific community—achieve good names by virtue of long histories of responsible reporting, quality articles, and exhaustive peer review.

In 2001, the Davis team laid its findings before the world. It did so not in a sports journal, not in a physiology journal, and not in a general-interest journal of good reputation, such as *Science*. Instead, it reported the yoga findings in *Preventive Cardiology*, the journal that Amsterdam had recently founded and on which he served as editor in chief.

In theory, his editorial control did nothing to diminish the study's credibility. The journal, after all, was peer reviewed. Amsterdam told me that the manuscript was sent to several reviewers with whom he had no relationship, making their evaluation "blind" and unbiased. Moreover, *Preventive Cardiology* was the official journal of the American Society for Preventive Cardiology, a professional group. Still, a situation in which the most important gatekeeper at a journal also submits his or her own work for publication can foster a perception of a conflict of interest. Did reviewers go easy on the manuscript to curry favor? Did the editor have a financial stake in the journal's success, and thus an incentive to make bold claims that would draw wide attention, raising the journal's readership?

A related problem centered on the sheer magnitude of Amsterdam's submissions. *Preventive Cardiology* carried so much of his own work that the journal, despite its professional affiliation, seemed less like an impartial forum than a vanity press. The same issue that featured the yoga study carried another one of his papers. In all, the quarterly journal that year published four of his articles. No other author came close. His work, except for the yoga study, focused on medical aspects of heart disease.

That led to a final topic of procedural significance—whether *Preventive Cardiology* was the right place for the yoga study. The Davis team reported a range of athletic findings, not just ones related to the heart. It seems like its natural home would have been an athletic forum, perhaps the *Journal of Exercise Physiology*. But the authors, for whatever reason or reasons, instead chose the pages of *Preventive Cardiology*.

The study came across as strong and authoritative. For instance, the Davis scientists reported that the fledgling yogis racked up solid gains in muscular strength. One test centered on knee extension—the act of straightening out the leg while raising a heavy weight. On average, the students improved 28 percent. They also showed greater flexibility. On average, they increased the amount they could bend forward (as in the Frog) by 14 percent. Their backward stretches (as in the Sun Salutation) improved even more, rising on average nearly 200 percent.

Unfortunately, the gains in aerobic conditioning—the primary interest of Amsterdam the cardiologist—were quite small. Even so, the Davis scientists judged them to be statistically significant. They reported that VO_2 max rose on average 7 percent. Moreover, they judged that the positive finding stood out from all previous studies, marking a milestone in the scientific evaluation of yoga.

"The present study," the authors declared, "is the first to show improvements in cardiorespiratory endurance by direct measurements." The scientists concluded that the overall results of their study indicated that Hatha yoga "would meet the objectives of current recommendations to improve physical fitness and health."

That was a big claim for what was indisputably a small investigation—for what its authors conceded was a "pilot study" that amounted to a preliminary look in search of noteworthy trends. The scientists offered no comment on how the small observed gain in aerobic conditioning measured up to the official recommendations of such groups as the American College of Sports Medicine, although their use of the conditional tense, "would meet," bespoke caution.

Nor did the authors put the aerobic figure into a wider context. They made no comparison of the 7 percent rise to what a sedentary individual might gain from endurance training, where scientists had found that peak oxygenation could increase up to 50 percent.

"In summary," the Davis scientists said, "the results of this investigation indicate that eight weeks of Hatha yoga practice can significantly improve multiple health-related aspects of physical fitness."

It was, arguably, a small step for the recognition of yoga as an aerobic activity—a step grounded in the discipline's growing incorporation of such vigorous poses as the Frog and the Sun Salutation. Or perhaps it was simply a fluke. The lack of experimental controls increased the chance of false readings.

Whatever the study's scientific merit, the leaders of the yoga community, long on the defensive when it came to cardiovascular issues, seized on the modest finding as a breakthrough. It was hard proof, they asserted, that yoga is all an individual needs to stay fit. The contention was a bold restatement of Gune's early claims. Only now—in theory, at least—it had the steel of modern science.

A portrait of the aerobics research formed the heart of a 2002 article in *Yoga Journal*. The glossy magazine prides itself on giving readers "the most current scientific information available." It spread its lengthy cover story on yoga fitness over nine pages and illustrated it with lots of color photographs of yogis in scientific labs undergoing close scrutiny. A main location for the documentary photos was the University of California at Davis. In its article, *Yoga Journal* reported that it had carefully surveyed the world of science and discovered solid evidence that "optimal fitness" requires no running or swimming to strengthen the heart and no weight lifting to build the muscles.

"Yoga is all you need," it declared, "for a fit mind and body."

The article said nothing about the downbeat findings of Cooper and the Duke scientists. It did, however, highlight the Davis study, calling the 7 percent rise "a very respectable increase" and hailing the aerobic finding as a breakthrough. Even so, the article, like the Davis authors, provided little context for the figure—making no comparison, for instance, to what endurance training can do for peak oxygenation.

Reaching further, *Yoga Journal* filled its article with profiles, testimonials, and anecdotal studies of people who hailed the yoga-alone perspective.

It quoted Dina Amsterdam. "I haven't done anything but yoga and some hiking for ten years," she said. "Yoga completely brought me back to physical and emotional health."

The Davis and *Yoga Journal* articles quickly became the go-to authorities around the globe for demonstrating that yoga alone was vigorous enough to meet the aerobic recommendations. The door had opened a crack, and a blast of aggressive marketing shot through.

One of the flashiest promoters was YogaFit, a commercial style that originated in Los Angeles. Its founding goal was to make yoga an integral part of the fitness industry. The style combined push-ups, sit-ups, squats, and other repetitive exercises with traditional yoga postures in a flowing kind of Vinyasa format. A centerpiece was the Sun Salutation. Seeking a wide audience, the style hailed sweat over what it characterized as yogic mumbo-jumbo, focusing instead on earthy rewards. For instance, its YogaButt program claimed to "totally transform your thighs and glutes," resulting in a bottom that is "sleek and sexy." YogaFit sold. Starting in the 1990s, its fast workouts spread through gyms, spas, and health clubs, with thousands of women taking up the contemporary hybrid. Its big hit was YogaButt. To satisfy demand, the company developed a course of training that could certify instructors in four hours.

YogaFit presented itself as a plunge into extreme fitness. Beth Shaw, its founder, claimed that the vigorous style focused minds, trimmed fat, toned bodies, and provided "a tough cardiovascular workout." Her promotional literature, when enumerating the fitness benefits of the style, cited the number one payoff as "cardiovascular endurance."

In 2003, the company sought to substantiate her cardio declarations. The sixteen-page paper, "Health Benefits of Hatha Yoga," cited no lab studies that YogaFit had sponsored. Instead, it reviewed the existing research. The paper cited the Davis study, the *Yoga Journal* article, and other inquiries as demonstrating that the style offered a serious path to the heights of cardiovascular fitness.

As usual in such tellings, the paper ignored the negative findings and the context. Still, it made the best of a tenuous situation and called Yoga-Fit and other energetic styles of yoga "aerobically challenging."

The good news spread. It traveled far beyond the insular world of yoga into mainstream culture. There, amid the blur of health and beauty tips, it got promoted as a scientific insight—with all the weightiness that such a discovery implied.

In 2004, *Shape*, which calls itself the lifestyle magazine for the active woman, hailed the Davis findings as proving that yoga provided all the cardiovascular benefits that anyone could want. "You don't need traditional cardio," it assured its readers, which it put at more than six million. The attainment of this most challenging of fitness goals, the magazine added, requires "nothing more than a yoga mat."

A principal dynamic in the psychology of scientific advance is the action–reaction cycle. Its workings are often on public display in the case of big claims, especially when the perception arises that the claimants have offered inadequate evidence to back up their declarations. At that point, the pendulum starts to swing in the opposite direction and the organized skepticism of science takes over. Rivals seek to poke holes in the original claim and try to discredit the original arguments. At times, the resulting disputes get settled quickly. But sometimes they drag on for decades as each side seeks to assemble evidence weighty enough to settle the argument once and for all.

Yoga's claims of aerobic excellence got caught up in that kind of reactive cycle. A large assertion had been made and had received considerable public notice—that yoga alone is sufficient to achieve cardiovascular fitness.

The claim was big and so were the stakes. If true, yoga could enter the pantheon of activities that global authorities had identified as vigorous enough to produce the array of cardio benefits—to raise stamina and lower the risk of heart disease, diabetes, cancer, and many other diseases.

From a business angle, the claim was pure gold. It could turn a simple form of exercise requiring no costly equipment or investment into a dazzling profit center. The pronouncement caught the attention not only of supporters but, increasingly, of skeptics.

The wave of scientific reaction started in 2005 even as the aerobic claims continued to echo and multiply through yogic and popular culture. It began at Texas State University. Carolyn C. Clay, a young scientist who practiced yoga, talked four colleagues into joining the investigation. Their study appeared in The *Journal of Strength and Conditioning Research*, the scientific forum of the National Strength and Conditioning Association, a nonprofit group of scientists and athletic professionals. The researchers looked at twenty-six women. That was more than twice as many subjects as in the Davis investigation. Moreover, the scientists examined the women not only as they did yoga but as they walked briskly on treadmills and rested in chairs. That gave the scientists a reasonable basis for comparison. It was an experimental control meant to enhance the reliability and—not inconsequentially—the credibility of their measurements.

Another precaution centered on skill. The scientists recruited volunteers from a university yoga class, and the subjects had practiced for at least a month. The experience factor implied that the moves and postures would be more precise and rigorous than with beginners, in theory strengthening the aerobic stimulus. It bespoke an effort to take the measure of yoga as regular exercise.

Clay and her team also brought new precision to the measurement of oxygen intake. Unlike the before-and-after methods of the Duke and Davis studies, the Texas researchers fitted their subjects with face masks hooked up to breath analyzers, producing direct readings of respiration. The scientists judged that the gains in accuracy would outweigh any inconvenience.

The yoga session was shorter than in the Davis study. It lasted just a half hour, compared to an hour and a half. The scientists said they designed it to resemble a routine in a health club. The Texas study, like the Davis investigation, put Sun Salutations at the heart of the session.

The investigators cited the Davis paper in reviewing prior research. But their findings bore little resemblance. Perhaps most conspicuously, the Texas scientists explicitly addressed how their findings measured up to the official recommendations.

The team examined a variation of VO_2 max known as maximum oxygen uptake reserve. It expresses the difference between oxygen consumption at peak levels of exercise and during rest. Since the resting metabolic rate of individuals can vary, exercise physiologists consider the reserve formula a more accurate way of making comparisons of athletic fitness. (The method is similar to how the vital index took personal factors into account.) The American College of Sports Medicine, in promoting aerobic conditioning, recommends that individuals draw on 50 to 85 percent of their maximum reserve. By contrast, the Texas scientists found that women walking briskly on the treadmill used about 45 percent.

And yoga? The women, while doing the routine, achieved far less— only 15 percent. The results, the scientists reported, "indicate that the metabolic intensity of hatha yoga is well below that required for improving cardiovascular health."

The only encouraging news centered on the Sun Salutation. Clay and her team said the fluid pose turned out to represent the workout's most aerobic aspect—a wide belief in the yoga community that had previously gone untested. The scientists found that Sun Salutations drew on 34 percent of the maximum reserve—more than twice the overall yoga session. And they suggested that the reading, though "significantly lower" than the 50 percent minimum of the American College of Sports Medicine, was nonetheless high enough for yoga teachers to consider putting more emphasis on the vigorous pose.

"To increase intensity," the researchers said, "it appears that the Sun Salutation or similar series of *asanas* should comprise the greatest portion of a Hatha yoga session."

Another downbeat finding emerged in 2005, just a month later. The study was done at the University of Wisconsin. It centered on thirty-four women with no yoga experience and no history of regular exercise. The women were divided into yoga and control groups. The yogis did fifty-five minutes of Hatha three times a week for two months while the non-yoga group did no exercise at all. Compared to the Texas study, the workout was longer and presumably more vigorous.

The investigators in Wisconsin found gains in strength, endurance, balance, and flexibility. But not in VO_2 max. "The intensity just wasn't there," noted John Porcari, one of the scientists.

The Wisconsin team did a companion study to see if Power Yoga—a demanding series of poses based on the Ashtanga system, with emphasis on flowing postures like the Sun Salutation—posed a greater aerobic challenge. The scientists recruited fifteen participants with at least intermediate experience. It turned out that the heightened vigor did make a difference, but only slightly. "You certainly sweat," Porcari said. "But it's not an aerobic workout."

He disagreed with the Texans on the idea of introducing wide customizations meant to increase yoga's vigor. Porcari said that adding more energetic postures as a way to boost cardio benefits would, by definition, come at the expense of flexibility, balance, and the other traditional benefits.

"It's always a trade-off," he said. "Yoga was designed for relaxation, primarily. The more aerobic you make yoga, the less improvement you'll see in those other areas."

Many yoga studies go unnoticed. The Wisconsin inquiry made waves, probably because its sponsor was the American Council on Exercise, a nonprofit group that seeks to protect consumers from risky and ineffective fitness programs. That gave the study added authority and exposure. The council, based in San Diego, published a digest of the Wisconsin study in its magazine and sent out a press release. The statement noted that the Wisconsin scientists had found that each Hatha session burned just 144 calories—similar to a slow walk.

"Aerobics?" *The Washington Post* asked in its headline. "Not Among Yoga's Strengths."

Yoga Journal took notice—defensively, acting like a true believer in denial. Its headline said it all: "Flexible *and* Fit."

The magazine faulted the Wisconsin study, as well as the reaction of the news media, and went on to cite new evidence of yoga's aerobic benefits. Once again, it found support at the University of California at Davis—the main source of its original good news on VO_2 max some four years earlier. A Davis researcher, *Yoga Journal* reported, had studied four yoga instructors who displayed levels of fitness comparable to someone who jogged three or four times a week. The news media, the article insisted, had fallen for a misleading story and had missed an inspiring one.

But the new evidence was thin. *Yoga Journal* gave no details about the new Davis study, just the claims. And, as it turns out, the study was never published. Its existence amounted to a rumor, although the readers of *Yoga Journal* could be forgiven for thinking otherwise.

The more obvious problem was that the Davis scientist had drawn a comparison between extensive and small efforts—comparing teachers who did "several hours of yoga a day" to joggers who ran as little as three times a week. The finding implied that running was far more aerobic—just what an impartial observer might conclude.

"I think you just proved the point," a reader wrote *Yoga Journal* in noting the lopsided comparison.

A final study, published in 2007, sought to settle the debate once and for all. It fairly breathed thoroughness and rigor. For instance, it did its recruiting in the studios of Manhattan, where youth, fierce competition, and starry clientele had resulted in challenging routines and gifted students—some of the best the planet had to offer. The sites ranged from downtown, to Midtown, to the Upper West Side. They included the torture chambers of Bikram Yoga ("we forge bodies and minds of steel"), the stylish removes of Levitate Yoga ("be free to wear the latest Louis Vuitton or Prada items"), and the sunny halls of the World Yoga Center ("created with a pioneering and idealistic spirit").

The researchers came from the Brooklyn campus of Long Island

University as well as the Mailman School of Public Health of Columbia University, a star of the biological sciences. They knew their stuff. The lead scientist, Marshall Hagins, had a doctorate in biomechanics and ergonomics and a clinical doctorate in physical therapy, and had practiced yoga for a decade.

The study's funding signaled its gravitas. Often, yoga investigators list no source of financial backing in their published work, implying that they undertook it on their own or with the aid of anonymous colleagues. That was the case with the Davis study. Such research tends to be modest in scope because the funding tends to be modest. Not so mainstream science. There, investigators typically go out of their way to thank their patrons—in the life sciences, often federal agencies. So it was with the New York study. The team in its published report said it had received support from the National Institutes of Health, the world's premier organization for health-care research.

The New York team recruited twenty subjects who had practiced yoga for at least one year, felt comfortable doing Sun Salutations, and could perform such advanced poses as the Headstand. The group consisted of two men and eighteen women.

The scientists judged that some of the previous studies had significant flaws. For instance, subjects were inexperienced or had been forced to wear clumsy masks and mouthpieces. "Such techniques," the scientists noted, "may alter the performance of the yoga activities and therefore provide invalid estimates." *Invalid estimates.* In the polite world of scientific discourse, that was tantamount to ridicule.

Seeking better results, the scientists made their measurements while the subjects did yoga in a special chamber that could track overall changes in respiratory activity. It let the yogis move about freely even while being scrutinized intimately. Known as a metabolic chamber, the rare and costly piece of scientific equipment was located at Saint Luke's–Roosevelt Hospital on the city's Upper West Side, near the Columbia campus. It was, in effect, an airtight cell. Machines linked to the metabolic chamber could measure a subject's exact consumption of oxygen, exhalation of carbon dioxide, and radiation of metabolic heat. Columbia scientists often used the chamber to study obesity. They would examine a subject's metabolic rate during meals, sleep, and light activities. But now they lent their apparatus to the scrutiny of yoga. The aim was not to track the addition of layers of fat but to see how efficiently yoga burned calories by fanning the body's metabolic flames. In terms of sophistication and accuracy, the chamber was light-years away from the rough bags that Hill had strapped on his runners, and from the traditional sets of before-and-after measurements that some modern scientists had used to track VO_2 max. It was cutting-edge.

The architects of the New York inquiry, as with most scientists who study humans, made sure its design included the imposition of controls meant to increase the likelihood that any observed changes were real rather than false clues or statistical flukes. Thus, the subjects in the metabolic chamber engaged sequentially in three different activities—reading a book, doing yoga, and walking on a treadmill. To provide a more detailed basis for comparison, the scientists had the subjects walk on the treadmill at different speeds. The imposed rates were two miles per hour and three miles per hour, the latter a fairly vigorous pace.

The yoga was pure Ashtanga, the brisk, fluid style descended from Krishnamacharya. The workout began with twenty-eight minutes of Sun Salutations followed by some twenty minutes of standing poses such as the Triangle and Padahastasana, a forward bend in which the student grabs the feet and brings the head down to the knees. It ended with eight minutes of relaxation in the Lotus position and the Corpse pose. Overall, the yoga session lasted nearly an hour. Its strong focus on Sun Salutations made the routine one of the most vigorous to undergo careful examination.

Hands to Feet, *Padahastasana*

Despite the added zing, the scientists concluded that the yoga session failed to meet the minimal aerobic recommendations of the world's health bodies. Its oxygen demands, they reported, "represent low levels of physical activity" similar to walking on the treadmill at a slow pace or taking a leisurely stroll.

The only glimmer of cardiovascular hope centered, once again, on Sun Salutations. The New Yorkers found the oxygen challenge of the pose "significantly higher" than for the slow treadmill. A practice incorporating Sun Salutations for at least ten minutes, they wrote, may "improve cardio-respiratory fitness in unfit or sedentary individuals." But the flip side, the scientists added, was that the posture offered few heart benefits for seasoned practitioners.

Seeking a wide context, the scientists were careful to note that other research had demonstrated that yoga aided the body and mind in ways that extended far beyond athletics and aerobics. Hagins, the lead researcher, remarked in a university news release that the discipline had "positive health benefits on blood pressure, osteoporosis, stress and depression." Yoga, he added, "may convey its primary benefits in ways unrelated to metabolic expenditure and increased heart rate."

That kind of prudence became the standard view in the world of science. Decades of uncertainty ended as a consensus emerged that yoga did much for the body and mind but little or nothing for aerobic conditioning. The California study receded as an anomaly, and the investigations in Texas, Wisconsin, and New York got cited repeatedly and respectfully in the scientific literature. Science and its social mechanisms had assessed a big claim and found it wanting.

In 2010, a review paper documented the new accord. In the halls of science, the review article is a hallowed tradition because of its pithy generalizations. It gives a critical evaluation of the published work in a particular area of research and then draws conclusions about what is legitimate progress and what is not, what is good and what is bad. By definition, it weighs *all* the evidence. Given the rapid advance of science around the globe, as well as the soaring numbers of reports, such analyses are seen as increasingly important to upholding the standard of wide comprehension. Whole journals do nothing but publish review articles.

The 2010 paper examined more than eighty studies that compared yoga and regular exercise. The analysis, by health specialists at the University of Maryland, found that yoga equaled or surpassed exercise in such things as improving balance, reducing fatigue, decreasing anxiety, cutting stress, lifting moods, improving sleep, reducing pain, lowering cholesterol, and more generally in raising the quality of life for yogis, both socially and on the job. The benefits were similar to those that had surprised the Duke team.

In summary, the specialists reported that yoga excelled in dozens of examined areas.

But the scientists also spoke of a conspicuous limitation for an activity that had long billed itself as a path to physical superiority. The authors noted that the benefits ran through all the categories—"except those involving physical fitness."

For the world of science, the issue seemed to be settled. But for the world at large? In truth, all the labors and analyses got little or no attention from the public, and most certainly not from yogis. The one exception was the Wisconsin study, which made a few waves. Overall, the rest of the studies sank without a ripple.

The lack of public reaction was especially notable in the yogic community. In theory, it was the main beneficiary of the findings. Even so, no yoga book to my knowledge reviewed the developments or commented on the implications. No guru expounded on the details. Bikram Choudhury offered no barbed rejoinders. *Yoga Journal* fired off no more rebuttals. The disregard persisted even though the lead researchers in the Wisconsin and New York studies were themselves yogis who could be seen as sympathetic to the discipline.

In the early days of modern yoga—in the era of Gune and Iyengar—science had been a role model. Influential gurus paid attention. No more. The affair was over.

Some individuals and authors who did yoga—or who knew the diversity of modern athletics—seemed to understand the substance of the new findings. But they tended to be exceptions. In popular culture, yoga went on its merry way, oblivious to the conclusion of science, believing deeply in its aerobic powers, often selling itself as superior to sports and exercise as the one and only way to attain that most fashionable of goals—ultimate fitness.

Yoga Journal continued its claims, hailing vigorous Hatha in 2008 as "a good cardio workout."

The vibrant cover of *Yoga for Dummies* publicized its two authors as holding doctoral degrees—the ultimate academic credential. Their experience seemed to magnify their authority. The cover identified one man as the "author of more than 40 books," and the other as an "internationally renowned Yoga teacher." In the book's second edition, published in early 2010, the two authorities hailed the Sun Salutation for its "aerobic benefits." More generally, they assured readers, the newer, more vigorous formulations of the ancient discipline let practitioners "work up a sweat" to achieve "aerobic-type" workouts.

Even *The New York Times* lost its bearings. One of its companies, About.com, addressed a frequent question of readers, "Does Yoga Keep You Fit?" Yes, came the unequivocal answer. Ann Pizer, the website's "Yoga Guide," said recent science had revealed that students doing yoga more than twice a week "need not supplement their fitness regimes with other types of exercise in order to stay very physically fit." She cited the original *Yoga Journal* article, the one that had turned the Davis study into a publicist's dream come true.

An indication of the fog's deep penetration was how it crept into the pages of *Hotel Management and Operations*—an industry guide in its fifth edition. In 2010, it advised readers that vigorous styles such as Ashtanga were ideal for getting a "cardio workout." The language wasn't as strong as Beth Shaw's "tough cardiovascular workout." But it nonetheless put the practice right up there with the sweaty rigors of fast treadmills.

The Davis and *Yoga Journal* articles kept getting cited, their claims immortalized on the Internet. The Huffington Post ran a link to *Yoga Journal*'s glowing tribute.

The aerobic mythology sped across cyberspace until it found its way to HealthCentral.com, a flourishing commercial site that sells drugs and gives away medical advice. The site claims more than seventeen million readers a month and, on its home page, proudly declares that it features material from Harvard Health Publications, the arm of the Harvard

Medical School that seeks to publicize the most authoritative health information. The site's reference to Harvard included a picture of the school's crest—a shield bearing three open books, their pages spelling out *veritas*—"truth" in Latin.

"Does Yoga Provide Enough of a Cardio Workout?" HealthCentral .com asked its readers in 2010.

The answer came from never-never land, despite the implication that the Harvard Medical School had somehow been involved in producing or vetting the information. It came from a place where the ancient practice had somehow morphed into a modern fitness machine.

"Rest assured," the site told its readers. "Yoga is all you need."

In recent years, many people have learned to ignore the exaggerated claims and the unabashed gurus (with or without fleets of Rolls-Royces). They lift weights to build muscles and run to challenge their hearts, even while pursuing yoga for flexibility and its other rewards. They are known as cross trainers. Their diverse workouts complement one another to produce a balance of benefits.

Cross training has no gurus, no schools, no fees, and no advocacy groups. Even so, it has a growing number of adherents, including top athletes. For instance, Alan Jaeger, a baseball pro with a passion for yoga, runs a school in Los Angeles for big-league pitchers. His stars run, stretch, meditate, listen to music, do yoga poses, and meditate again— doing that kind of routine for hours before getting around to throwing a ball.

The popularity of cross training serves as an instructive counterpoint to yoga's overstated fitness claims. It speaks to the wisdom of people who pay close attention to what their bodies tell them—and to the recommendations of public health officials.

The guidelines for aerobic exercise developed slowly over the course of the twentieth century, as we have seen. Ultimately, the enterprise drew on the work of hundreds of scientists. A less conspicuous effort got under way during the same period. It focused not on the straightforward goal of quantifying oxygen uptake and determining its role in physical fitness but on the more difficult and amorphous problem of understanding how changing situations can shape human emotion.

By nature, the mood question was far more challenging than studying the ups and downs of VO^2 max. An added complication was that the investigation of human feeling had less political traction because it involved fewer scientists, received less funding, and had less institutional support than was the case with research into sports and physical training. The pool of candidates shrank further still when the focus was on such curiosities as yoga. The discoveries in that case could be quite serendipitous in nature, as with the Duke investigators.

Scattered groups of scientists nonetheless made notable progress in the course of the twentieth century. In recent years, their work has become more abundant and substantial. The field—after an early history of false starts and digressions—now appears to be coming into its own.

The trend is significant. In the end, it may elucidate one of yoga's most important benefits.

MOODS

S at Bir Khalsa chatted amiably as we walked down the street. His beard was long and gray, his turban white, and his bracelet made of steel—all signs of his Sikh religion. He was not, however, Indian. Born in Toronto of European stock, he had converted to Sikhism decades ago upon taking up Kundalini Yoga, an energetic form that emphasizes rapid breathing and deep meditation.

No one on Longwood Avenue seemed to give the turbaned Sikh a second look. Boston that spring day was gorgeous. An early shower had scrubbed the air, leaving it awash in sunlight. Flowers and trees were blooming. Men and women were shedding their coats. People fairly hummed along the sidewalks.

We had just eaten lunch at Bertucci's, a bustling restaurant where Khalsa had finished his meal with *bomba*—"the bomb" in Italian. The dessert consisted of balls of vanilla and chocolate gelato dipped in chocolate and covered with almonds, whipped cream, and chocolate sauce. I could see why his kids loved the place.

Maybe it was the sugar high, or the beautiful day, or the yoga. Whatever the cause, the air fairly pulsed as Khalsa—a faculty member at the Harvard Medical School and one of the world's leading authorities on the science of yoga—laid out his findings and ambitions. The friendly man of fifty-six turned out to have a lot.

At Harvard, Khalsa had pursued a bold program of research that explored how yoga can soothe physically and emotionally. His focus was practical—and structured that way deliberately to demonstrate yoga's social value. He had examined how its powers of unwinding can promote sleep and ease performance anxiety among musicians, and was now organizing a study to see if its calming influence could help high school students better fight the blues and everyday stress. Khalsa had ten yoga investigations in various stages of development.

With energy and articulate zeal, he described his research as a way to help yoga break from its fringy past and go mainstream.

"What ever happened to mental hygiene?" he asked rhetorically. "It doesn't exist—and never did. When you went through high school, you were never taught how to deal with stress, how to deal with trauma, how to deal with tension and anxiety—with the whole list of mood impairments. There's no preventive maintenance. We know how to prevent cavities. But we don't teach children how to be resilient, how to cope with stress on a daily basis.

"There's a disconnect," he continued. "We've done dental hygiene but not mental hygiene. So the question is, 'How do we go from where we are now to where we need to be?'"

Khalsa argued that the only way to convince people about the value of yoga and establish a social consensus that encouraged wide practice was to conduct a thorough program of scientific research. He added that recommendations for regular toothbrushing had started that way and illustrated the potential value of good yoga studies.

"That's my mission in life," Khalsa told me.

This chapter examines not only Khalsa's research but many inquiries into how yoga can lift moods and refresh the human spirit. It starts with the earliest research and ends with the most recent. The arc of the narrative is really a detective story. The studies began with the muscles (and how yoga can relax them), went on to study the blood (and how yoga breathing can reset its chemical balance), and eventually zeroed in on the subtleties of the nervous system (and how yoga poses can fine-tune its status). The discipline was found to lift and lower not only emotions but also their underlying constituents—the metabolism and the nervous system.

The mood benefits detailed here are very real, unlike some of the aerobic claims of the last chapter. But the field also has its popular myths. They tend to be outright errors, probably rooted in ignorance rather than subtle shadings of the truth done with profit in mind.

Psychologists tell us that a fundamental building block of emotional life is strong feeling, such as fury or affection. By definition, moods are considered less intense, more general, longer lasting, and less likely to arise from a particular stimulus. They are seen as drawn-out emotions. For instance, joy over a period of time produces a happy mood. Sadness over time results in depression. Unlike sharp feelings such as rage or surprise, moods tend to last for hours and days, if not weeks. If intense, they can color our life perceptions—at times dramatically.

Moods are central to meaning in life and thus, in the judgment of psychologists, more important than money, status, and even personal relationships because they affect the happiness quotient that we assign to life activities. As the saying goes, a rich man in a bad mood can feel destitute, and a poor man in a good mood rich beyond words. To a surprising degree, moods define our being.

It turns out that the word arose in the earliest days of the English language and that its first definitions resonate with existential import. "Mood" was originally a synonym for "mind." In Old English, the word *mod* meant "heart," "spirit," or "courage."

An intriguing question that investigators have yet to address is whether yoga can change an individual's pattern of moods—in other words, a person's core emotional outlook. Can the regular practice of Sun Salutations produce a sunny disposition? Does yoga bring about what might be considered characteristic states of affability?

Many people have looked to their own experience on such matters and found that, overall, yoga lifts their emotional life. Significantly, the vast majority are women.

The conventional wisdom is that episodes of major depression strike women twice as frequently as men. The drug evidence is stark. A survey found women nearly three times as likely as men to take antidepressants—with usage as high as one in every four or five women.

If those characterizations are right, yoga should resonate strongly with women as a way of fighting the blues. I personally saw evidence of that attraction. In the early winter of 2010, I joined dozens of women (and a few guys) who had gathered to learn about using the discipline as a means of emotional uplift.

"It really saved my life," Amy Weintraub told us during her introductory talk. "I wouldn't be here." It was Friday night at Kripalu, the yoga center in the Berkshires of western Massachusetts. Weintraub, author of *Yoga for Depression*, was leading a weekend seminar on mood management.

She came to her calling after a life of crippling dejection and numbness. "I moved as through a fog," she recalled in her book. "I lost keys, gloves, and once, even my car." Antidepressants did little. Then she found yoga. The fog lifted. In a year, she was off drugs and soon became a yoga instructor. Her rebirth came with deep feelings of emotional strength.

At Kripalu, for three days, Weintraub marshaled every available weapon in the yoga arsenal to teach us how to seize control.

"You'll be feeling lighter and brighter or your money back," she said with a smile. Her methods were not particularly strenuous. But they all took aim with great precision at lifting the spirit. We relaxed. We visualized. We did balancing poses that forced us to shift our attention from mind chatter to the here and now. We laughed. We stretched. We made calming sounds. We did Breath of Joy—inhaling, bringing our arms slowly up to the sky, then exhaling with a breathy "Haaaa" while bringing our arms down rapidly. By the end of it, we glowed, lit from within.

Weintraub understood the science and told our class about a number of studies and researchers. It turned out that she knew Khalsa and was working with him on one of his investigations—to compare the benefits of yoga with those of psychotherapy. She also had a book in the works: *Yoga Skills for Therapists.*

All this may sound new and fresh. But it turns out that Harvard, Boston, and Massachusetts have long played host to individuals and institutions with abiding interests in yoga's emotional sway. In fact, I suspect that is the core attraction of Kripalu, which has drawn waves of interest for decades and describes itself as the nation's largest residential center for yoga and holistic health. The facility is located on hundreds of rural acres far from the usual pleasures and distractions of urban life.

On the Friday that I signed in—with the leaves of the trees gone and the area dappled in white from a recent snowfall—Kripalu succeeded in registering nearly five hundred guests for its weekend classes. The vast majority were women.

Thoreau characterized yogis as having no earthly care: "Free in this world as the birds in the air." In 1849, he told a friend that he considered himself a practitioner—the first known instance of a Westerner making that claim. "I would fain practice the yoga faithfully," Thoreau wrote. "To some extent, and at rare intervals, even I am a yogi."

At Harvard, his alma mater, William James looked favorably on yoga as a means of mental regeneration. The famous psychologist, trained as a medical doctor, zeroed in on one of yoga's most basic exercises—the simple but systematic relaxation of the muscles.

The pose is called Savasana, from *sava*, the Sanskrit word for "dead body." Today we call it the Corpse. It is the easiest of yoga's positions. Rather than twisting or stretching, students simply lie on their backs, eyes shut, and let their arms, legs, and other body parts go limp. In this state of repose, students relax their muscles as much as possible, entering a condition of deep rest. It is usually done at the end of a yoga class and seems to have been around for centuries.

Corpse, *Savasana*

James, in his 1902 book, *The Varieties of Religious Experience*, identified that kind of letting go as "regeneration by relaxation" and suggested that it could not only revitalize the spirit but advance the more ambitious goal of fostering healthy life attitudes. "Relaxation," he wrote, "should be now the rule."

A graduate student paid close attention. His name was Edmund Jacobson. A physiologist, he had come from Chicago to work on a doctoral degree at Harvard. The gospel of relaxation caught his eye and, following the lead of James and other professors, he threw himself into an experimental study. It focused on the startle reaction—in particular to how subjects reacted when a strip of wood was slapped down on a desk with a sudden *crack*. To his surprise, Jacobson found that relaxed subjects had no obvious reaction. He surmised that deep relaxation caused mental activity to drop.

Jacobson tried relaxation himself. Like many students, he suffered bouts of insomnia. But in 1908 he taught himself how to relax and found that the lessening of muscular tension let him enjoy a good night's rest.

Jacobson became a convert. Upon taking a job at the University of Chicago, he pursued an ambitious agenda of research and treatment, more or less founding the medical field of trained relaxation. His books included *Progressive Relaxation* (1929) and *You Must Relax* (1934), which went through more than a dozen printings and editions.

Jacobson would have patients close their eyes and tense and relax a body part, concentrating on the contrast. In time, they would get the hang of reducing the tension. Jacobson claimed that his method produced remarkable cures, eliminating everything from headaches and insomnia to stuttering and depression.

To satisfy his own curiosity—and to convince skeptics of the method's importance—Jacobson worked hard to gather a body of objective evidence. His goal was to develop a machine that would let him track tiny electrical signals in the muscles and measure subtle currents of a millionth of a volt or less. In his efforts, Jacobson got considerable aid from Bell Telephone Laboratories—then the world's premier organization for industrial research, which in time won a half-dozen Nobel Prizes. The collaboration between medical doctor and industrial giant resulted in innovations that foreshadowed the electromyograph, a medical instrument that records the electrical waves of skeletal muscles.

Jacobson made what are regarded as the world's first accurate measurements of tonus—the normal state of slight muscular tension that aids posture and readies the body for action. The instruments showed that his methods could induce deep calm.

One patient was a woman who suffered a skull fracture when she got caught in a folding bed. For years afterward, she complained of a nervous condition that made her overly emotional. When Jacobson hooked her up to his apparatus, he found that her muscles did in fact exhibit electrical spikes even when she tried to remain as calm and tranquil as possible. Later, after the woman learned to relax deeply, the waves vanished. And she found new strength and stability in her emotional life.

As Jacobson studied one way of letting go, other means of calming the mind and refreshing the spirit came under investigation. One scientist looked at controlled breathing and did so through the lens of one of the most fashionable emerging fields of the day, psychology.

• • •

Kovoor T. Behanan was born and raised in India. In 1923, he graduated from the University of Calcutta with distinction. He traveled to the United States and Yale, first to study religion, then philosophy and psychology. In 1931, he won a fellowship to travel back to India and evaluate the psychology of yoga.

Behanan did the reasonable thing. He journeyed to the world capital of yoga research—Gune's ashram in the mountains south of Bombay. There he threw himself into learning yoga under Gune's personal guidance. From April 1932 to March 1933, Behanan practiced every day, doing postures, breathing exercises, and concentration training. He then returned to Yale determined to do a series of scientific experiments that would explain his newfound joy.

Behanan started in early 1935, after he had been practicing yoga for more than three years. He studied his own mental reactions, seeing the work as exploratory. In particular, he zeroed in on one of yoga's easiest kinds of breath control.

Ujjayi Pranayama is known as Victorious Breath. Despite its intimidating name, the style involves simply breathing in and out with great deliberation, the glottis in the throat slightly contracted so the breath makes a hissing sound, like the soft roar of the ocean. Behanan would inhale slowly through both nostrils. After filling his lungs to the brim, he would hold the breath for the same length of time as the inhalation, and then exhale steadily for the same duration.

Resting adults breathe anywhere from ten to twenty times a minute. Ujjayi is *much* slower. Behanan reports that he did the exercise at a rate of twenty-eight cycles in twenty-two minutes, or a little more than one per minute. In other words, he breathed about ten times slower than a resting adult. It was the kind of slowdown that Paul had described in his book *A Treatise on the Yoga Philosophy.*

The psychological testing went on for thirty-six days. Behanan would do the evaluations before and after the breathing exercises to see how they changed his state of mind. The tests consisted of adding numbers, breaking codes, identifying colors, doing puzzles, and performing little exercises in physical coordination.

He published his results in a 1937 book, *Yoga: A Scientific Evaluation.* He now had a doctorate from Yale, as the title page noted prominently, and the book was well received. *Life* ran a formal portrait of Behanan in coat and tie and photographs of half-naked students upending themselves in tricky poses. The feature was spread over two pages. *Time* ran a glowing review. It called him handsome, thirty-five, and a first-class poker player.

Behanan's central finding turned out to be a lucid confirmation of Jacobson's surmise about deep relaxation causing a drop in mental activity. Across the board, the breathing exercise brought about what Behanan called "a retardation of mental functions." The finding, he conceded, might leave readers a little surprised.

All the tests took him longer to complete, up to twenty-six seconds longer. The yoga breathing had its greatest impact—and produced the greatest lag—on his math abilities.

His findings, Behanan noted, contradicted the popular image of yoga as a magic elixir that endowed its practitioners with superhuman powers. But he hailed the mental slowing as important because of its repercussions for mood. The breathing exercise, Behanan reported, brought about a state of deep relaxation that produced "an extremely pleasant feeling of quietude." The inner pleasure became even greater if he added concentration exercises. "I would like to prolong it indefinitely," he wrote of the floating state, "if it were in my power to do so."

The evidence indicated that the mental slowing was temporary and Behanan held out the possibility that the period of refreshment might actually produce an overall improvement in "our normal intellectual faculties."

At the book's end, Behanan summarized his own reaction to his newfound discipline. Yoga, he said, had remade him.

Before, he had frequent headaches, felt run down, and lacked what he called pep. But his days at the ashram gave him new energy and "emotional stability." Behanan saw the same in his ashram colleagues.

"They were the happiest personalities that I have known," he recalled. "Their serenity was contagious."

If only he had stopped there. In his book, Behanan went on to describe experiments he had done at Yale in respiratory physiology (a very different field with very different methods and measurement techniques that in many respects are more difficult). He reported that Ujjayi caused a spike in oxygen consumption—more than any other breathing style that he investigated. To all appearances, the oxygen boost seemed to be the secret of yoga serenity.

But it was only the myth, yet again. Behanan had left India before Gune cast doubt on the popular doctrine of the surges, and the research of the Yale investigator proved to be flawed.

Unfortunately, his false report added to the unshakable durability of the oxygen myth, which still haunts popular yoga. And his misapprehension neglected what turns out to be one of yoga's main sources of leverage over the unruly currents of human emotion—the manipulation of the body's relationship with carbon dioxide, which both Paul and Gune had begun to glimpse.

Today, after many decades of research, it is fairly easy to distinguish between fact and fiction, despite the complexities of respiratory physiology. The big picture is the best place to start.

The atmosphere of our planet is 21 percent oxygen. That's a lot. In comparison, the levels of carbon dioxide are five hundred times smaller. The human body exploits this ocean by means of hemoglobin—the remarkable protein inside our red blood cells that soaks up oxygen like a sponge and carries it from the lungs to the tissues. Typically, the refreshed hemoglobin of a resting person is nearly saturated with oxygen, holding virtually as much as it possibly can. The usual figure for the level of absorption is 97 percent.

For yoga, the glut of oxygen in the air and the saturation of hemoglobin in the lungs mean that fast or slow breathing does little to change the levels that enter the bloodstream—as Gune found at his ashram and as I found at the University of Wisconsin. The vital gas is available in large quantities no matter what.

The body's consumption of oxygen does go up and down. But science demonstrates that it does so in response to changes of muscle activity, metabolism, and heart rate—not breathing styles. As we saw in the last chapter, cardiovascular fitness can raise peak oxygen consumption.

The story with carbon dioxide in the bloodstream is dramatically different and variable. Consider a person breathing in a relaxed way. Fresh air mixes in the lungs with stale air, creating an inner environment where carbon dioxide levels remain fairly high. This person, in typical fashion, ventilates the lungs so inefficiently that each relaxed inhalation replaces less than 10 percent of the gas.

Now consider what happens if that individual starts to breath fast. Blasts of fresh air with extraordinarily low concentrations of carbon dioxide (three-hundredths of 1 percent of the atmosphere) rush into the lungs, lowering the inner levels. Nature seeks to equalize the concentrations. So diffusion quickly draws more carbon dioxide out of the bloodstream and into the lungs. The result is that the body's levels plunge.

This view is anything but contentious. It is the standard account found in hundreds of medical textbooks as well as official pronouncements of the U.S. Navy. Its responsibilities for thousands of professional divers make it a global authority on human respiration. Fast breathing "lowers body stores of carbon dioxide," the Navy's diving manual states, "without significantly increasing oxygen stores."

The common name for fast breathing is hyperventilation, and the common danger is passing out. It can also result in dizziness, headaches, light-headedness, slurred speech, and numbness or tingling in the lips, hands, and feet. The drop in carbon dioxide influences mood in many ways. One is through respiratory alkalosis. It heightens the excitability of nerves and muscles—so much so that many circuits short out, producing tingling in the hands and spasms in the muscles.

Yogis often feel such sensations after doing many rounds of Bhastrika, a fast-breathing exercise known as Breath of Fire or Bellows Breath, after its Sanskrit definition. Bhastrika puts the emphasis on exhalation rather than inhalation, like the bellows of a blacksmith. In *Light on Pranayama*, Iyengar says the repeated blasts of the exercise create "a feeling of exhilaration."

While exhilaration and nerve excitement go up, fast breathing does something else that has critical repercussions for mood, mental outlook, and potentially health—it robs the brain of oxygen.

The reason is that the drop in carbon dioxide causes blood vessels in the brain to contract, reducing the flow of oxygen and producing light-headedness and perhaps blurred vision. Other symptoms include dizziness and giddiness. In extreme cases, a person can hallucinate or pass out.

What this means in plain English—as crazy as it sounds, as counterintuitive as it seems, as contrary to the teachings of popular yoga as it appears—is that fast breathing lowers the flow of oxygen to the brain, and does so dramatically. Scientists have found that it cuts levels roughly in half. That plunge is why people faint.

The misunderstandings about fast breathing, once stark, have grown more subtle over the years and decades. Yoga authorities who look into the science of respiration now tend to contradict one another on whether it can disrupt the carbon dioxide metabolism and result in harm. Some issue stern warnings about hyperventilation and caution new students to take Bhastrika in small increments so the body can adjust gradually to the challenge. Others argue that fast breathing—if done right, especially with methods they advertise as superior—does nothing to upset the carbon balance.

The evidence of the classroom suggests that fast breathing can indeed pose a threat. Many a beginning student of yoga has fainted, and even intermediate students can feel dizzy or pass out. The science on the topic is rather limited. It does, however, suggest that advanced students can adapt to the respiratory push or learn to do Bhastrika and other varieties of fast breathing in ways that lessen the dangers.

In 1983, three scientists in Sweden reported on a study of three highly trained yogis who did "high frequency breathing" for anywhere from thirty minutes to an hour. The scientists began the experiment by fitting the subjects with catheters to ease the sampling of blood. The arterial tap was meant to open a window on how the changing atmosphere of the lungs was affecting the rejuvenation of the blood and the release of carbon dioxide. The scientists took blood samples as the yogis rested—to create a basis for comparison—and after they had performed fast breathing for at least ten minutes.

The results showed relatively modest drops—4 percent, 11 percent, and 30 percent. Interestingly, the most experienced of the yogis, a man who taught the other two as well as regular classes in breath control, exhibited the intermediate decline. His was the decrease of 11 percent rather than 4 percent. That suggested that factors other than proficiency and experience in pranayama could limit the release of carbon dioxide. Overall, the scientists reported that none of the advanced yogis developed symptoms of respiratory alkalosis—the blood disturbance that can result in light-headedness and collapse.

The overall repercussions for mood and respiratory physiology are radically different if the yoga breathing is slow rather than fast. The process starts with such varieties as Ujjayi—the kind that Behanan did. He was breathing about ten times slower than a resting adult.

The consequences again center on carbon dioxide—only this time its rise in the bloodstream, not its fall. Modern investigations echo Paul's studies of more than a century ago. Today, a standard figure is that cutting lung ventilation in half prompts blood levels of carbon dioxide to double. And the ensuing dilation of cerebral blood vessels means the brain now gets more oxygen, not less.

Slow breathing turns out to have deep mental ramifications, with increases in calm alertness and raw awareness. Behanan called the state "an extremely pleasant feeling of quietude."

Scientists who study animal behavior have linked slow breathing to heightened vigilance. When an animal is ready to protect itself, its exhalations slow. Its heart rate tends to fall. The animal carefully assesses its surroundings to see if it can relax or needs to flee or fight.

As Paul suggested, many aspects of yoga reinforce slow breathing and limit the exhalation of carbon dioxide, including the repetition of mantras and chants. In 2001, Luciano Bernardi, a medical internist at the University of Pavia, in Italy, reported on a study of nearly two dozen adults. His team found that the repetition of a mantra cut the normal rate of respiration by about half, reinforcing mental calm and producing an enhanced sense of well-being.

In short, science over the decades has learned a lot about how yoga breathing affects a person's mood and underlying metabolic state. Fast styles tend to excite and slow ones to calm.

And it has *nothing* to do with getting more or less oxygen into the practitioner's body, contrary to innumerable yogis and gurus, video discs and yoga books, blogs and newsletters. Nevertheless, some yoga authorities go so far as to issue delusional warnings.

"You're not used to so much oxygen pouring into your system!" Choudhury, the guru of hot yoga, cautions students who do his deep breathing and begin to feel dizzy. He says nothing of big carbon dioxide drops, the real cause of blackouts.

The blunders go on and on. Breath of Fire "increases oxygen delivery to the brain," said *Kundalini Yoga*, richly illustrated and highly accessible to beginners. Actually, as we just saw, it does the exact opposite, dramatically so.

The Complete Idiot's Guide to Yoga praises the discipline's breathing as "one of the best things you can do to keep your body filled with oxygen." The advertised effects sounded a lot like the calm serenity that high carbon dioxide levels can induce.

The confusion about yoga breathing as a way to fill the body and brain with oxygen goes far beyond simple misstatements and their dissemination through countless books, articles, and videos. Of late, it has spread to a whole new style of the discipline. *Oxygen Yoga* promotes itself as beneficial "for anyone in need of more oxygen." Its authors have a line of books. The newest—*Oxygen Yoga: A Spa Universe*, published in 2010—recommends that health resorts adopt the style for "added revenue."

It is said that every disaster has a silver lining. In a similar way, the failure of yoga investigators to find miracle workers who could stop their hearts and live without air led to a major advance in understanding the brain. And that discovery in turn revealed one of the most important ways that yoga can sway emotion. It happened in the decades after Behanan did his experiments, from roughly the 1940s to 1970s.

The lesser discoveries ended up revising a major tenet of the medical world—that the human body has two nervous systems that are entirely distinct. The newer one starts in the outer brain and radiates out in the nerves that let us move our skeletal muscles and go about our daily lives. The older one begins in the lower brain and regulates the internal muscles, the organs, the instincts, and other primal functions. It is called the autonomic nervous system.

The medical credo of the day held that its activities were automatic and, with notable exceptions (such as breathing), beyond the control of the conscious mind. But scientists who studied gifted yogis kept documenting abilities that contradicted this tidy picture. In study after study, they found that yogis could seize control of autonomic functions and make dramatic changes of body activity. The automatic system, it turned out, contained options for all kinds of manual overrides.

One of the scientists was Thérèse Brosse, a French cardiologist who examined Krishnamacharya. She and her colleagues wrote extensively on how advanced yogis could unwind in surprising ways, slowing the heart rate and blood flow. Another was Bagchi. Despite his campaign to expose yogic miracles as false, he documented wide yogic control over autonomic functions once considered beyond reach. A 1957 paper of his found "an extreme slowing" of such fundamentals as respiration and heart rate. He concluded that overall, yoga brings about "deep relaxation of the autonomic nervous system."

The star of autonomic control was an Indian yogi named Swami Rama. Among other things, laboratory studies showed that he could use his mind alone to change the temperature of his hand, creating a gap of up to eleven degrees across his palm.

The autonomic system is bifurcated, and the studies showed that advanced yogis could seize control of either side. The sympathetic side promotes the body's fight-or-flight response, inhibiting digestion and moving blood to the muscles for quick action. It does so partly by telling the adrenal glands to squirt out adrenaline, a natural stimulant that speeds up body functions. Early biologists called it "sympathetic" because they saw its functions as acting in concert or *sympathy* with one another—all at once. The other side is known as the parasympathetic. It governs the body's rest-and-digest functions, calming the nerves, promoting the absorption of food, and curbing the flow of adrenaline.

The sympathetic system is the body's accelerator, and the parasympathetic the brake. Working together, the two manage the body's overall energy flow, one preparing for its expenditure, the other for its conservation. For instance, the sympathetic system raises the heart rate, and the parasympathetic slows it down.

The two also wield control over human moods and emotions rooted in primal energy states—the ups and the downs, the exhilarations of Iyengar and the quietudes of Behanan. The inner states resonate with some of the most fundamental of all human emotions—whether individuals feel safe and protected or threatened and endangered. They reflect not only our survival instincts but the mood swings of childhood.

The investigations showed that yogis had a special talent for applying the brake. Their adroit slowing of the metabolism and related functions was especially impressive in that it overcame a strong evolutionary bias. The demands of survival mean that the body, left to its own devices, always favors the accelerator. After all, the sympathetic nervous system is essentially a means of emergency response and easily aroused, keeping the body ready for battle or retreat, awash in adrenaline.

Yogis have pressed these pedals for ages. But recently, a new breed of practitioner has arisen who can not only work the pedals but draw on a wealth of contemporary science to explain the process and how it relates to the experiences of regular practitioners.

I speak of Mel Robin.

Many yoga teachers have a reverential aura. Not Robin, not on a rainy Saturday in Pennsylvania as he strode into a crowded yoga studio, his manner relaxed, his beard trim. New Age music played softly in the background.

"My name is Mel and I wrote this music," he quipped, getting a laugh.

Someone asked if he performed it, too.

"Don't get smart," he replied. And with that, Robin shattered the usual atmosphere.

The Yoga Loft of Bethlehem sat atop an old brick building that seemed to date from the city's steel days. It had bare wooden floors and large windows. The studio had put aside its regular Saturday classes for a special program featuring Robin on "The Science of Inversions"—poses in which the feet or torso go above the head. The ads suggested that students bring along a copy of his *Physiological Handbook for Teachers of Yogasana* or buy one at the studio's shop.

Women in their thirties tended to dominate the room. Several were yoga instructors, as the class was advanced. Robin was seventy-three and, in typical yoga style, failed to look or act his age.

He sat on the bare wooden floor, stretching and talking, and over the next three hours took us through some of the nuances of autonomic control. His focus that day was not the kind of mental gymnastics that Rama had displayed. Rather, it was explaining how normal poses can result in autonomic shifts.

Headstands, he said, tend to excite the fight-or-flight response, especially in nervous beginners. By contrast, he added, the Shoulder Stand pressed the parasympathetic brake, soothing the spirit and making it "one of the most relaxing postures in yoga."

We paired off and Robin, a teacher in the Iyengar style, toured the room helping twosomes do Shoulder Stands. When he got to me, I asked if more was known about the reasons for the relaxation.

Robin said the pose calmed because it seized control of one of the most important functions of the autonomic system—the regulation of blood pressure.

It is well known that good health depends on the pressure staying in a narrow range. If it drops too low, the brain gets insufficient blood and we get dizzy, weak, and faint. In extreme cases, organs can fail, producing such breakdowns as cardiac arrest. High blood pressure has its own hazards, though long-term rather than immediate. It stresses the heart and arterial walls, producing hypertension. This is a risk factor for stroke, heart attack, and kidney failure. Because of such dangers, the human body over the ages has evolved a striking array of sensors and defense mechanisms that constantly take pressure readings of the blood vessels and make suitable adjustments.

Robin said the Shoulder Stand tweaked one particular kind of sensor. It lay in the carotids—the major arteries that run through the front of the neck carrying blood to the brain. The carotid sensors make sure the brain gets the right amount of blood and, given the brain's importance, get serious attention. Sensors embedded in the arterial walls monitor bulging or contracting that indicate changes in blood pressure.

Shoulder Stand, *Sarvangasana*

But in the Shoulder Stand, Robin said, the chin presses deeply into the neck and upper chest, clamping down on the carotids and making the local pressure very high. That rings alarm bells and the parasympathetic brake flies into action. It assumes that the delicate tissues of the brain are reeling from too much blood and orders the heart and the circulatory system to compensate with pressure cuts. The main response signals go through the vagus—the large nerve that starts in the brain stem and wanders among the lungs, heart, stomach, and other abdominal organs.

Robin clapped his hands to illustrate the urgent nature of the parasympathetic commands.

"Don't pump so often. Don't pump so hard. Open the diameters. Vasodilate"—the term for vessel relaxation that allows blood to flow at a more leisurely pace.

I thanked him. Gune may have recommended the Shoulder Stand to Gandhi for its calming effects, but he had nothing on Robin in explaining how it worked.

Robin volunteered that the scientific approach to understanding the poses made some yogis uncomfortable.

"There are people who say, 'You've crossed the line. That's not yoga. Look at Patanjali. There's nothing about the workings of the sympathetic nervous system.'

"They are very, very traditional people," he continued. "My own feeling is—I agree. It's not yoga. It's about yoga and understanding it, and that lets you do better yoga."

And with that, Robin turned his attention to other students.

Better yoga. The phrase echoed in my head. A few days later, I called Yoga Loft and signed up for a series of four courses that Robin planned to teach on the science of yoga. The last focused entirely on the autonomic nervous system.

Right off, he cast the topic in a new light. Most portrayals stress psychological factors—such as existential threats that prompt the fight-or-flight response, or peaceful interludes that bring about rest-and-digest states of contentment. But Robin said the systems could be stimulated not only by environmental factors but from the bottom up by conscious actions. An example, he said, centered on the muscles.

"If you're frightened, your muscles get tense," he said. "But if you do muscle work, that also excites the sympathetic nervous system." It was a fascinating observation that had all kinds of implications for life and explaining the influence of yoga postures.

Robin pointed to a place in his book that listed which parts of the body came under autonomic control. The table, spread over four pages, described more than one hundred functions—everything from sleep and gastric secretions to vasoconstriction and shivering. Each entry was followed by a reference number, or several reference numbers, that pointed to the book's end section of scientific reports. The table seemed to represent a labor of love that summarized decades of research.

Robin had us turn to another page that listed nine unusually potent effects of sympathetic stimulation. They include a quicker heartbeat (to prepare for action), dilated pupils (to admit more light for better attention to potential threats), and changed blood chemistry (to stimulate clotting in case of bleeding). Most people have no conscious control over such autonomic responses. But two items on the list stood out as relatively easy to influence—muscle tone and respiration rate. Robin called them keys to the hidden world of autonomic control.

We practiced poses that worked the muscles, seeking to excite the sympathetic nerves. "Any kind of exercise, any kind of muscular work" will do it, Robin told us. He added that the same held true for respiration. "Anything you do to speed up your breath will speed up most parts of the sympathetic system."

It was a fascinating idea. His activity rule, given the list of autonomic functions we had just looked at, suggested that a disciplined individual could gain leverage over dozens of the body's most important and inconspicuous functions. His rule also suggested that different Hatha styles had very different autonomic effects. For instance, Ashtanga, with its fast, flowing movements and its emphasis on Sun Salutations, works the muscles a lot and would thus stimulate the sympathetic system. In contrast, Iyengar, with its emphasis on static poses, would seem to lend itself to parasympathetic dominance.

Robin expanded the activity explanation a step further and showed us subtle ways in which a pose could engage the parasympathetic brake. His ideas were an elaboration on what he had told me during the inversions class.

He focused on the heart itself. Robin noted that the right atrium—the upper chamber that gets blood from the veins—bears a sensor that gauges its fullness. When pressure is low, he said, the sensor signals the heart to beat faster, increasing the blood flow. When pressure is high, the heart slows down.

Robin said inversions worked beautifully—as with carotid pressure—to fool the heart into slowing. It happened because upending the body dramatically increased the flow to the right atrium. Normally, gravity helped a little between the head and the heart. But turning the body upside down let gravity work over a much larger area, strengthening venous flow from the feet, legs, and torso.

"It's all downhill," Robin noted. "So the heart overfills."

The rising pressure in the right atrium then signaled the heart to beat slower. That signal, Robin noted, also caused the heart to reduce the strength of its contractions. It was a one-two punch. Overall, the atrial mechanism showed yet another way in which yoga could work inconspicuously to reset the metabolism.

Robin had us do a heart test. First, we monitored our pulses, measuring their rate against the sweep of a watch. Then he had us move to a wall, lie on our backs, and raise our legs in a relaxed state of partial inversion. Once again, we measured. He noted that the beat was probably slower (which I found to be the case).

A good yoga practice, Robin said toward the end of class, involved poses that cycled through the accelerator and the brake so the autonomic system got a thorough workout. Robin said the resulting realization of energetic flexibility over the usual conditions of metabolic life resulted in new abilities to achieve states of inner balance and harmony.

"A large part of the benefits," he said, "results from going through a couple of cycles every time we practice."

The clues gathered over the decades about yoga's repercussions for human emotion first began to come together in a significant way at Harvard. Herbert Benson was a physician eager to ease Western tension with regular doses of Eastern calming. He and his colleagues studied the issue at the university's medical school, examining the effects of meditation, yoga, and other soothing practices. Benson called his insight *The Relaxation Response*. His book, published in 1975, sold more than four million copies and became a modern classic on undoing stress.

Benson found that simple techniques could have dramatic repercussions on his subjects, cutting their heart rate, respiratory rate, oxygen consumption, and blood pressure (if elevated to begin with). Overall, he and his colleagues showed that relaxed practitioners entered a state known as hypometabolism—a wakeful cousin of sleep that exhibits low energy expenditures. He called the relaxation response "an inducible physiologic state of quietude" that healed and revitalized.

After Benson, many scientists sought to expand his findings and zero in on particular disciplines, including yoga.

Mayasandra S. Chaya was an Indian physiologist in Bangalore who had practiced yoga since she was ten. Chaya led a team that studied more than one hundred men and women. The scientists prescribed a diverse Hatha routine sure to press both the metabolic brake and accelerator. The dozen poses included the Triangle (Trikonasana), the Shoulder Stand (Sarvangasana), the Locust (Shalabhasana), the Cobra (Bhujangasana), the Bow (Dhanurasana), and the Thunderbolt (Vajrasana), as well as fast and slow breathing techniques. At a session's end, the subjects assumed the Corpse (Savasana) for a period of conscious relaxation. The men and women—their average age thirty-three—followed the prescribed routine for at least a half year.

The scientists assessed how the routine affected the basal metabolic rate—the energy spent on the body's housekeeping functions. In a standard method, they measured the flow of respiratory gases—oxygen and carbon dioxide—as a way of gauging how bright the inner fires were burning, as measured in calories.

In 2006, Chaya and her team reported that regular yoga practice cut the basal metabolic rate by an average of 13 percent. The results were even more pronounced when broken down by sex. The men on average cut their resting energy by 8 percent. But the women achieved reductions of 18 percent—more than double the metabolic declines of their male counterparts.

It bespoke the wisdom of Robin's comments about autonomic cycles. The ups and downs worked to increase not only outer flexibility but inner suppleness as well, giving body and mind the freedom to sink into Benson's kind of quietude—to just be. It was a secret of letting go.

The metabolic dips also raised an issue that bore on personal appearance rather than moods but was too central for the scientists to ignore.

Chaya's team noted that physiological slowing of yoga in theory "creates a propensity for weight gain and fat deposition." In other words, individuals who took up the discipline would reduce their basal metabolic rate to such an extent that they required less food and fewer calories—or would add pounds if they ate and exercised in the customary manner.

This novel finding of physiology might have won attention, but it clashed with the happy talk of popular yoga. Teachers, the Internet, and how-to books had long echoed with confident declarations that yoga speeds up the metabolism and results in an almost magical loss of weight. It was one of modern yoga's credos—equal in some respects to the illusory surge of oxygen to the body and brain. In truth, the metabolic conviction was so deeply held that no inconspicuous finding in faraway India stood a chance of undoing the fashionable myth.

Tara Stiles exemplified the durability. The attractive model turned yoga teacher favored short shorts and tank tops, and managed to keep herself beanpole thin. In Manhattan, she ran Strala Yoga in NoHo, a chic neighborhood north of Houston Street. In 2010, she came out with *Slim Calm Sexy Yoga*, its cover emblazoned with a photograph of Stiles in an eye-catching pose. The book featured an endorsement from Jane Fonda and rose fast to become the number-one yoga seller on Amazon. Early in 2011, *The New York Times* profiled Stiles, saying the twenty-nine-year-old displayed not only sexy good looks but down-to-earth charm.

The title of her book led with *Slim*, and Stiles worked hard in the text to deliver on the promise. She devoted a chapter to slenderizing and explained what seemed to be the scientific basis for why the discipline worked so well at keeping off the pounds. Yoga, she declared, will "rev up your metabolism." Getting specific, she recommended a series of postures meant to throw the body into high gear in the service of shedding weight. "Even if you think you have a sluggish metabolism," Stiles said, "practicing this routine twice a week will keep it humming—and help you burn calories all day." She emphasized the point in large type spread across the top of the page, advertising her custom routine as "Metabolism Revving."

The bold declarations of Stiles not only contradicted the body slowdowns that the Bangalore team had documented. They also clashed with the particulars that Mel Robin had told our class. For instance, Stiles listed the Shoulder Stand as one of her metabolism lifters. In contrast,

Robin had described the inversion as "one of the most relaxing postures in yoga." And Gune, of course, had recommended the pose to Gandhi for its calming action.

Chaya, the physiologist in Bangalore who had practiced yoga since childhood, told me that the secret of weight loss had nothing to do with a fast metabolism and everything to do with the psychological repercussions of undoing stress. "Yoga affects the mind—and desire," she said. "So you eat less."

If yoga can foster serenity and lift moods, what are its repercussions for depression, where the requirements for emotional uplift are much greater? It was a tough question. Amy Weintraub in her book, *Yoga for Depression*, recounted her own experiences and prescribed many practical ways of dealing with the blues. But depression has many faces and, in its most severe forms, is crippling.

Everyday gloom involves low feelings and loss of pleasure, perhaps due to minor setbacks. That kind of dejection, by nature, is fleeting. By contrast, the symptoms of clinical depression last two weeks or more. A seriously depressed person can feel anything from hopelessness and discouragement to worthlessness and despair. The World Health Organization says that, every year, nearly one million despairing people take their own lives. That is more than the number of people killed annually in crimes or war. In industrialized societies, despite floods of antidepressants, rates of suicide and depression are going up, not down.

Once again, scientists in Boston zeroed in on the question. Their studies went far beyond clinical trials and patient evaluations to examine neurochemistry. The team represented the elite of the Boston medical world—the Boston University School of Medicine and the Harvard Medical School and its McLean psychiatric hospital. The hospital is famous for its neuroscience research as well as its extensive roster of celebrity patients, including the mathematician John Nash, the poet Sylvia Plath, and the musician James Taylor.

Chris C. Streeter, the head of the team, held faculty appointments in psychiatry and neurology at the Boston University School of Medicine and lectured in psychiatry at the Harvard Medical School. Plus, she knew yoga and knew people who knew yoga. Her team focused on an important chemical in the human brain that goes by the tongue-twisting name of gamma-aminobutyric acid—or, easier to say, GABA. It is a major neurotransmitter and regulator of the human nervous system. Many reports have linked depression to low GABA levels. So a smart question was whether yoga went about easing depression by raising concentrations of the neurotransmitter.

Scientists have known about GABA since the 1950s. But it took a long time to understand its role in the brain and to develop the scientific tools to easily track its comings and goings. GABA works by blocking actions rather than causing them. It is known as an antagonist. Such chemicals, when they bind to cellular receptor sites in the nervous system, disrupt interactions and inhibit the functions of other neurotransmitters. In general, GABA slows the firing of neurons, making them less excitable. So high levels of the neurotransmitter have a calming effect. When alcohol and drugs like Valium bind to GABA receptor sites, they increase the molecule's efficiency and thus promote its actions as a sedative and a muscle relaxant. GABA itself tends to promote relaxation and reduce anxiety.

By the 2000s, brain imaging had advanced to the point that tracking GABA could be done fairly inexpensively. Scientists judged the time right to address the yoga question.

The team found lots of potential subjects. Boston, starting with Thoreau and James, had evolved into a yoga hotspot. In modern times, it pulsed with many thousands of practitioners.

The team selected eight who practiced a number of diverse styles. They were Ashtanga (the gymnastic style developed by Pattabhi Jois, a student of Krishnamacharya's), Bikram (the hot yoga of Choudhury), Hatha (the ancient classic), Iyengar (the modern classic), Kripalu (developed by the Berkshires center), Kundalini (the heavy-breathing style popularized by Yogi Bhajan, a Sikh mystic), Power (an aggressive form of Ashtanga), and Vinyasa (a flowing style developed relatively late in life by Krishnamacharya and popularized by his student Srivatsa Ramaswami). The subjects had practiced yoga anywhere from two to ten years. They were all white, mostly female, mostly single, and averaged twenty-six years in age. Prior to the study, all had practiced yoga at least twice a week.

The team measured GABA levels before and after an hour-long yoga session. The routine was standardized to focus on asanas and related breathing. At the start and end, the students could engage in brief sessions of quiet contemplation. But they were allowed no extensive periods of meditation or pranayama. The study guidelines called for at least fifty-five minutes of common asanas, such as inversions and backbends, twists and Sun Salutations. To ensure a degree of standardization, a research staff member with yoga training observed the sessions. The scientists compared the eight yoga practitioners to a control group of eleven individuals who did no yoga but instead read magazines and popular fiction for an hour.

The results, published in 2007, fairly glowed. The scientists found that the brains of yoga practitioners showed an average GABA rise of 27 percent. By contrast, the comparison group experienced no change whatsoever. Moreover, the yoga practitioners with the most experience or who practiced the most during the week tended to have real GABA surges. For instance, the practitioner who had done yoga for a decade experienced a GABA rise of 47 percent. One participant who practiced yoga five times a week had an increase of 80 percent, the levels of the neurotransmitter almost doubling.

The scientists concluded that yoga showed much promise for treating anxiety and depression. Perry F. Renshaw, a senior author of the study and director of brain imaging at the McLean Hospital, noted with understatement that any proven therapy that is cheap, widely available, and shows no side effects has "clear public health advantages."

Encouraged, the team embarked on a new study. This time the scientists looked at nineteen subjects and a control group of fifteen people who walked for exercise, which was seen as having the same metabolic expenditure as yoga. The main subjects had no significant yoga experience. They learned the Iyengar style from scratch and practiced it for three months.

The findings were published in 2010. They showed that even beginning yogis experienced major rises in the neurotransmitter along with improved moods and lessened anxiety. The average GABA rise was less than in the previous study—13 percent versus 27 percent, or about half as much. Still, the new yogis did better than the walkers. And, judging from the evidence, they felt much better about themselves.

Significantly, one of the eleven coauthors of the study was Liz Owen, an Iyengar teacher who ran classes in the Boston suburbs of Cambridge and Arlington. Owen had no doctoral or medical degrees but knew a lot about using yoga to lift moods. "Relax your body," her website advised. "Nourish your soul."

During this same period, Khalsa worked hard on studies meant to see if mood adjustment could have demonstrable benefits for diverse careers and life stages. One centered on musicians. Khalsa did his investigation with teachers from Kripalu and focused the research on a renowned establishment just down the road from the Berkshires yoga center—Tanglewood, the summer home of the Boston Symphony Orchestra and its academy of advanced study for young musicians. The goal was to see if doing yoga could help the beginners overcome stage fright in general and, more specifically, perform better for the demanding audiences that came to Tanglewood for summer concerts.

In 2005, Khalsa and Stephen Cope from Kripalu recruited ten volunteers from Tanglewood's prestigious fellows program. The five men and five women were aged twenty-one to thirty, the average just over twenty-five. They included singers, as well as those who played the violin and viola, horn and cello. For two months, the ten volunteers underwent Kripalu training. The options included morning and afternoon sessions seven days a week, a weekly evening session, an early-morning meditation session, and vegetarian meals at Kripalu. The investigation also included ten fellows recruited as controls who had no yoga training.

The results, though not earthshaking, were encouraging, as Khalsa and Cope reported in their 2006 paper.

The study had assessed performance anxiety that the musicians felt in practice sessions, group settings, and solos. The yogis showed no difference from the control group in practice and group settings but did demonstrate a striking drop in performance anxiety during solos. That made sense, Khalsa and Cope noted. Research showed that such nervousness was low during practice, moderate in group settings, and high in solo performances. So the mood effects, they reasoned, would show up more during solos.

During my visit with Khalsa, we sat in his Harvard office and pored over the Tanglewood results on his computer. A yoga mat was rolled up under his desk. "There's no question the kids loved it," he said. "The control group had hardly any change. But look at the yoga groups. Yoga brings you into the moment. It brings a feeling of joy or energy with activity, a kind of mindfulness."

The results were so positive, Khalsa added, that Tanglewood asked for more. He and Kripalu responded with an expanded study. The young musicians who immersed themselves in yoga, meditation, and Kripalu numbered thirty. And it turned out that their two months of summer practice lifted moods even higher.

In 2009, Khalsa and colleagues reported that the yogi musicians, compared to a control group, showed strong evidence of not only less performance anxiety but significantly less anger, depression, and general anxiety and tension. They loved it, like their predecessors.

Moreover, the scientists tracked down the students a year after the summer program and asked if their lives had changed. Most reported that they had continued doing yoga and meditation, and all said the experience had improved their performance skills.

The portrait of yoga that emerges from decades of mood and metabolic studies is of a discipline that succeeds brilliantly at smoothing the ups and downs of emotional life. It uses relaxation, breathing, and postures to bring about an environment of inner bending and stretching. The actions echo, in a way, how yoga pushes the limbs into challenging new configurations. They promote inner flexibility. As Robin observed, a good workout involves repeatedly pressing the accelerator and brake. Ironically, the overall result is a smoother ride.

No studies have examined the most extreme consequences. But the current evidence seems to suggest that yoga can reduce despair and hopelessness to the point of saving lives. You cannot read Weintraub's book and learn the details of her turbulent past—cannot watch her doing Breath of Joy, her face lit from within—without feeling the positive force of life affirmation.

If science reveals that yoga can excel at emotional uplift, it also shows that the discipline has a downside. It can do great harm.

IV

RISK OF INJURY

t is no surprise that a field that prides itself on the routine performance of twists, contortions, and dramatic bends of the human body can do a lot of damage. In a similar vein, it makes sense that circus performers—including tumblers and acrobats—also suffer high rates of impairment, and that running, bicycling, and other vigorous sports can result in painful accidents. Even so, yoga injuries are unsettling because of the discipline's image as a path to exceptional health. Many people turn to yoga as a gentle alternative to exercises that leave them hurt or intimidated. The idea of damage also runs counter to yoga's reputation for healing and its promotion of superior levels of fitness and well-being. Few practitioners anticipate strokes and dislocations, dead nerves and ruptured lungs.

The good reputation of yoga rests in no small part on the public silence of the gurus. Their virtual ban on the word "injury" made the topic of blinding pain and physical damage almost as unmentionable as Hatha's origins. Gune made no allusion to injuries in *Yoga Mimansa* or his book *Asanas*. Indra Devi avoided the issue in *Forever Young*, as did Iyengar in *Light on Yoga*. Silence about injury or strong reassurances about yoga safety also prevailed in the how-to books of Swami Sivananda, K. Pattabhi Jois, and Bikram Choudhury. In general, the famous gurus tend to describe yoga as a nearly miraculous agent of renewal. As one, they imply or state explicitly that ages of practice have shown the discipline to be free of hidden danger.

"Real yoga is as safe as mother's milk," declared Swami Gitananda (1907–1993), a popular guru who made ten world tours and founded ashrams on multiple continents.

Modern physicians, on the other hand, have taken an almost malicious delight in recounting the self-inflicted wounds of yoga practitioners and warning of danger, doing so in dozens of reports. Perhaps they are jealous of the admiration accorded to yoga teachers and get a thrill out of challenging yoga's mystique. Some have gone so far as to condemn yoga as intrinsically unsafe. What takes the edge off some of this criticism—especially during its first appearance—is how it often revealed a lack of deep knowledge about the workings of yoga but nonetheless managed to strike a tone of icy condescension. Even so, the medical professionals lavished attention on yogis who stumbled into their offices and emergency rooms writhing in pain, and wrote up detailed clinical reports on the accidents and injuries.

Like stones cast into a pond, these disclosures produced waves of reaction that in time affected the practice of modern yoga and ultimately helped make it safer—albeit after considerable resistance. Initially, some yogis challenged the reports as biased and mean-spirited. Others, perhaps taken with the mother's milk argument, tried to ignore the criticism or shrug off the injuries as an inconspicuous cost of doing business.

In recent years, the best teachers have responded to the warnings with new sensitivity (and better insurance policies). They put safety first, caution their students to proceed with care, and reject the one-size-fits-all mentality of early styles and instructors.

To yoga's credit, a number of knowledgeable practitioners have recently stepped forward to confront the physical threats quite directly in articles, books, bibliographies, and—most recently—detailed surveys of yoga injuries. The activists are generally reformers who seek to raise awareness of the dangers and offer precautions. The surveys, which can be alarming, suggest that yoga's recent popularity has created a rush of inexperienced teachers. Ironically, it seems that idyllic vacation spots are particularly treacherous.

Robin is one of the reformers. His books feature lengthy addendums that detail some of the ways in which yoga can go wrong. They tell of paralyzed limbs, bulging eyeballs, damaged brains—among other varieties of destruction, some verging on the bizarre. The appendices reflect his careful reading of the medical literature. They portray a hidden world of major trauma as well as minor problems such as sprains and torn muscles, which turn out to be surprisingly common. In his Pennsylvania class, we practiced a number of precautions, especially on how to unburden the neck in the Headstand and Shoulder Stand.

As a group, the activists tend to be in closer alignment with the findings of science than yoga traditionalists. Just as Robin and his Iyengar colleagues have redesigned the Headstand, some of the reformers have focused on reinventing some of the most dangerous poses or advising students to drop them altogether.

Such reevaluations may go against yoga's timeless image. But as we have seen, yoga has proved itself quite flexible in adapting to the needs and desires of different ages. Today, the long silence of the gurus has given way to scientific inquiries that are nurturing new strategies for injury prevention. The reform movement is a happy case study in what can happen if yoga and science cooperate, even grudgingly. The inconspicuous wave of reinvention promises to benefit millions of students around the globe and, not insignificantly, to help modern yoga live up to its good reputation.

In my travels, I learned of an experienced yogi who was said to know the inside story on yoga injury. Prominent gurus had supposedly come to him for help in rehabilitation and recovery. One client was reported to have received a hip replacement before reentering the celebrity life. I decided to track him down.

Glenn Black had traveled to India, studied at Iyengar's school in Pune, and, like the ancient yogis, spent years in solitude. He ran yoga intensives in the jungles of Costa Rica. In New York City, for a decade, he studied with Shmuel Tatz, a Lithuanian who devised a unique method of physical therapy that he dispensed from offices above Carnegie Hall to actors, singers, dancers, musicians, composers, and television stars. Black had settled down in Rhinebeck, New York, on the Hudson River. Honored as a master teacher and anatomist, he often taught yoga at the nearby Omega Institute, a New Age emporium. Black had a devoted following drawn to his earthy, no-nonsense style. He also had an elite bodywork clientele that included celebrities. Of late, he was said to have narrowed his client list down to a handful of billionaires.

One day I noticed that Black was scheduled to teach a master class in Manhattan. I hesitated but was told that resolve was more important than skill. I arranged to talk with him afterward.

On a cold Saturday in early 2009, I made my way to Sankalpah (aim, will, determination) Yoga, a third-floor walkup on Fifth Avenue between

Twenty-Eighth and Twenty-Ninth. The room was filled with lean bodies, roughly half of the individuals said to be teachers.

The class was brutal. Black joked, walked around a lot, talked constantly, played jazz on the sound system, watched us like a hawk, and cajoled relentlessly. Beads of sweat turned into rivulets. He was highly demanding yet surprisingly gentle, having us do lots of stretching, limb movements, and pose holding but no inversions and few classical postures. His teaching was nothing like the regimented styles. Instead, he worked us from the inside out. His approach was almost freeform and it seemed as if he was making it up as he went along, switching gears every so often to better challenge the range of aptitudes in the room or to pull us back from what he perceived to be some kind of cliff. In so doing, he conveyed a sense of intelligent vitality.

Through it all, he urged us to concentrate and try to develop our sense of attention and awareness, especially to the risky thresholds of pain. "I make it as hard as possible," he told us. "It's up to you to make it easy on yourself."

Playfully, he rejected any doubts about his style. "Is this yoga?" he asked as we sweated through an extremely unyogalike pose. "It *is* if you're paying attention."

Black told us a grim story. In India, he said, a yogi from abroad had come to study at Iyengar's school and threw himself into a spinal twist. Black said he watched in astonishment as three of the man's ribs gave way—*pop, pop, pop.*

After class, I joined Black and his companion, Evelyn Weber, on a cab ride back to their hotel. They said they were both born in 1949 and were turning sixty. Both looked much younger. "I am certified in nothing," Black remarked at one point. "I have no degrees. All I have is a ton of experience."

Discreet and luxurious, Hotel Plaza Athénée was located on the tree-lined Upper East Side at 64th between Park and Madison, with sister hotels in Paris and Bangkok. We went up to their suite. Daylight flooded the rooms. Weber served nuts and tea as we talked of yoga safety. Black sat on a couch, relaxed but serious.

He was amazingly blunt. My encounters with yoga denial and evasion had left me unprepared for such outspokenness and sweeping visions of better safeguards. It was radical. If Black ran the world, he would have many people—including many celebrities of the yoga circuit—relinquish not just difficult poses but the discipline itself. Students as well as celebrated teachers injured themselves in droves, he argued, because most were completely unprepared for yoga's rigors.

Black said the vast majority—"99.9 percent"—have underlying physical weaknesses and problems that make serious injury all but inevitable. Instead of doing yoga, "they need to be doing a specific range of motions for articulation, for organ condition," he said. "Yoga in general is for people in good physical condition. Or it can be used therapeutically. It's controversial to say, but it really shouldn't be used for a general class. There's such a variety and range of possibilities. Everybody has a different problem."

Black said he worked hard at trying to recognize signs of danger and knowing when a student "shouldn't do something—the Shoulder Stand, the Headstand, or putting any weight on the cervical vertebrae."

I asked if he ever modified poses to make them safer.

"Constantly," he answered. Referring to our just-completed class, Black noted how we had done a standing pose where we had put our arms behind our backs, clenched our hands together, and stretched our arms up. "I could see people's faces crunching, so I said, 'Bend your elbows.'" It was, he said, a safety valve.

"To come to New York and do a class with people who have many problems, and say, 'Okay, we're going to do this sequence of poses today'—it just doesn't work." Instead, he said, all classes had to be tailored to the range of particular student abilities on that particular day.

Weber noted that she had been studying with Black for a decade and had never experienced the same class twice.

Black said his guiding principle in teaching yoga was to downplay the asanas and put the emphasis on awareness. "It's harder to teach," he said. "But the risk of not teaching it is very great. If you just teach people to do an asana without taking them into deeper states of realization, their asanas are always going to be a struggle."

The superstars of yoga were so addicted to celebrity that they often overlooked the message of awareness and paying close attention to their bodies and anatomical limits, Black said. He told of famous teachers coming to him for healing bodywork after suffering major traumas. "And when I say, 'Don't do yoga,' they look at me like I'm crazy. And I know if they continue, they wouldn't be able to take it."

He said yoga celebrities seemed to have a predisposition to engage in not only personal denial but social evasion. "A yogi I know was going to be interviewed by *Rolling Stone* and said, 'I don't want to talk about injuries.'"

I asked about the worst injuries he had seen, and Black rattled off a long list. He told of big-name yoga teachers doing the Downward Facing Dog so strenuously that they tore Achilles tendons. "It's ego," he said. "The whole point of yoga is to get rid of ego." He said he had seen some "pretty gruesome hips. One of the biggest teachers in America had zero movement in her hip joint. The socket had become so degenerated that she had to have a hip replacement."

Downward Facing Dog, *Adho Mukha Svanasana*

I asked if she still taught. "Oh, yeah. And there are other yoga teachers that have such bad backs that they have to lie down to teach. I'd be so embarrassed."

Black said that he had never injured himself or, as far as he knew, been responsible for harming any of his students in thirty-seven years of teaching. "People feel sensations, sure, and find limitations. But it's done with mindfulness, not just because they're pushing themselves. Today, many schools of yoga are just about pushing people."

He told of his students reporting back to him on the aggressive tactics of other instructors. "You can't believe what's going on—teachers jumping on people, pushing and pulling and saying, 'You should be able to do this by now.' It has to do with their egos."

Black also chided students who practiced yoga for the excitements of status and cachet. "They take a class to show off their Missoni T-shirt or their leotards," he said, scowling. I asked his opinion of *Yoga Journal*, which over the decades had gone from a geeky nonprofit published by the California Yoga Teachers Association to a glossy magazine filled with ads for sexy clothing, travel adventures, and miracle weight-loss drugs. He declined comment.

While many gurus and yogis over the decades had remained silent on the threat of injury, or had denied its existence, or had grudgingly made limited concessions of danger, Black insisted that the threat was now indigenous to the discipline and just waiting to strike. He argued that a number of factors had come together in modern times to heighten the risk.

The biggest was the changing nature of students. The poor Indians of yoga's past normally squatted and sat cross-legged, the poses thus being in some respects an outgrowth of their daily lives. Now yoga had become a Western fad, swelling its unskilled ranks. Urbanites who sat in chairs all day now wanted to be weekend warriors despite their inflexibility and physical problems. Amateurish teachers ruled like drill sergeants and pushed cookie-cutter agendas. Such factors became all the more deadly, Black argued, with the distractions of modern vanity, which kept students and teachers from focusing on the importance of the here and the now, from listening to their bodies and understanding when they were about to cross the line from a wholesome stretch to excruciating harm.

The result was an epidemic.

"There has to be a degree of seriousness and dedication," he said. "Otherwise, you're going to get hurt."

The first scientific light on the topic of yoga injury fell decades ago. The reports appeared in some of the world's most respected journals—including *Neurology*, the *British Medical Journal*, and *The Journal of the American Medical Association*. The high-level debut signaled that the medical establishment saw the findings as important information that practicing doctors needed to know if they were going to help patients.

The reports began to emerge in the late 1960s, soon after the West had become newly interested in yoga.

A number of early findings centered on nerve damage. The problems ranged from the relatively mild to permanent disabilities that left students unable to walk without aid. For instance, a male college student had done yoga for more than a year when he intensified his practice by sitting upright for long periods on his heels in a kneeling position known as Vajrasana. In Sankrit, *vajra* means "thunderbolt." The position, also called the kneeling pose, is sometimes recommended for meditation. The young man did the pose for hours a day, usually while chanting for world peace. Soon he was experiencing difficulty walking, running, and climbing stairs.

Thunderbolt, *Vajrasana*

In Manhattan, an examination showed that both of his feet drooped because of a lack of leg control, and doctors traced the problem to an unresponsive nerve. It was a peripheral branch of the sciatic, the longest nerve of the body, which runs from the lower spine, through the buttocks, and down the legs. The damaged branch ran below the knee, normally providing the lower leg, foot, and toes with sensation and movement. Apparently, the young man's kneeling in Vajrasana had clamped his knees tight enough and long enough to cut the flow of blood to the lower leg, depriving the nerve of oxygen. The result was nerve deadening.

It was suggested that the young man simply give up the pose. Reluctantly he did so, opting instead to do his chanting while standing. He improved rapidly, and a checkup two months after the initial visit showed no lingering problems. In describing the case, the attending physician called the condition "yoga foot drop." The name stuck. In time, a number of similar cases emerged.

One of the worst featured a woman of forty-two. She fell asleep in Paschimottanasana—the Seated Forward Bend, its Sanskrit name meaning "stretch of the West." Upon awaking, she found her legs numb and weak. A medical team at the University of Washington, writing in *The Neurologist,* told of finding injuries to both her sciatic nerves that had crippled her legs. The scientists reported that the woman regained "some sensation" after three months of therapy but still displayed persistent foot drop.

Seated Forward Bend, *Paschimottanasana*

A half year after the mishap, the woman was still unable to walk without assistance. Her doctors said evidence of permanent nerve damage left them doubtful that she would ever recover full use of her legs.

If the first reported cases were relatively minor, a second wave soon emerged in which the consequences were little short of devastating. The reason was that the damage centered on the brain itself—not some peripheral organ or physiological subsystem. The news got worse. The blows to the body's most important organ arose not from stretching too much or holding postures too long but from the skilled practice of poses that practitioners did routinely and tended to see as completely safe.

The situation was so ominous that a leading British physician issued a public alert. In the conservative world of medicine, it is a rare day when abstract theorization comes ahead of clinical reports. Usually it is the other way around—first observation, then efforts at explanation and generalization. But the physician had the requisite stature to issue a sharp warning even before his peers had published any reports that described particular cases.

At the time, in 1972, W. Ritchie Russell was an elder statesman of British medicine. The string of acronyms after his name bespoke his status: M.D. (Medical Doctor), C.B.E. (Commander of the British Empire), F.R.C.P. (Fellow of the Royal College of Physicians), and D.Sc. (Doctor of Science). A neurophysiologist, he had distinguished himself in a long career at Oxford University that showed, among other things, that brain injuries could arise not only from direct impacts to the head but from quick movements of the neck as well, including whiplash. He published his pioneering research in the early 1940s as war swept Europe and neck injuries grew rapidly in number.

His new warning centered on how some yoga postures threatened to reduce the blood flow to the brain and cause the cerebral disasters known as strokes. The second most important cause of death in the Western world, right after heart disease, strokes often strike older people whose arteries get clogged with fatty deposits. The risk of dying from them rises with age. In addition, Russell worried about a fairly rare type of stroke that tended to strike relatively young, healthy people.

The word "stroke" is a euphemism for a range of destructive nastiness that develops when the regular flow of blood to the human brain gets interrupted. In many cases, the symptoms arise on just one side of the body because the brain's functional areas mirror the body's bilateral symmetry. Most strokes start as simple blockages. The flow of blood through an artery gets reduced or blocked entirely by deposits of fat, clots of coagulated blood, or the swollen linings of torn or damaged vessels, robbing the brain of oxygen. By definition, strokes traumatize and kill brain cells, which are known as neurons. A renewed flow of blood can sometimes mend beleaguered cells. And over time, nearby neurons can sometimes replace the function of dead cells. But damage can also be permanent. Stroke victims thus experience disabilities that range from passing weakness to lasting neurologic damage to death if the destruction involves vital brain centers. (Fast treatment can limit the damage, which is why health professionals urge speedy evaluations of suspected stroke victims, preferably within sixty minutes.) The symptoms of stroke vary widely because of the brain's highly specialized anatomy. For instance, conscious thought and intelligence arise in the outer layers of the brain, so strokes in those areas can affect speech and critical thinking.

Russell's concern went deeper. He worried about the inner brain, in particular a functionally diverse region toward the rear. His concern was that yoga postures that involved extreme bending of the neck might compromise the region's blood supply, destroying parts of the brain rich in primal responsibilities.

The human neck is made of seven cervical vertebrae that anatomists have numbered, top to bottom, C1 through C7. Their special shapes and compliant disks make the neck the most flexible part of the spinal column. Scientists have measured the neck's normal range of motion and found the movements to be extraordinarily wide. The neck can stretch backward 75 degrees, forward 40 degrees, and sideways 45 degrees, and can rotate on its axis about 50 degrees. Yoga practitioners typically move the vertebrae *much* farther. For instance, an intermediate student can easily turn his or her neck 90 degrees—nearly twice the normal rotation.

Russell had long specialized in understanding how the bending of the neck could endanger the flow of blood from the heart to the brain. His concern focused mainly on the vertebral arteries. By nature, every tug, pull, and twist of the head rearranges these highly elastic vessels. But major activity outside their normal range of motion can put them in jeopardy in part because of their unusual structure.

In traversing the neck, the vertebral arteries go through a bony labyrinth that is quite unlike anything else in the body and quite different from the soft, easy path that the carotids follow to the brain. The sides of each vertebra bulge outward to form loops of bone, and the arteries penetrate these loops successively in moving upward. The left and right vertebral arteries enter this gauntlet at C6 and run straight through the loops until they reach the top of the neck, at which point they start to zig and zag back and forth as they move toward the skull. Between C2 and

C1, they usually bend forward, and then, upon exiting the bony rings of C1, usually curve sharply backward toward the foramen magnum—the large hole at the base of the skull that acts as a conduit for not only blood vessels but nerves, ligaments, and the spinal cord. Anatomists describe the final journey of the vertebral arteries toward the brain as serpentine and report much variability in the exact route from person to person. It is not unusual for the tops of the vertebral arteries to branch out in a tangle of coils, kinks, and loops.

From decades of clinical practice and laboratory study, Russell knew that extreme motions of the head and neck could wound these remarkable arteries, producing clots, swelling, constriction, and havoc downstream in the brain. The victims could be quite young. His ultimate worry centered on the basilar artery. Located just inside the foramen magnum, the vessel arises from the union of the two vertebral arteries and forms a wide conduit at the base of the brain that feeds such structures as the pons (which plays a role in respiration), the cerebellum (which coordinates the muscles), the occipital lobe of the outer brain (which turns eye impulses into images), and the thalamus (which relays sensory messages to the outer brain and the hypothalamus and its vigilance area). In short, the basilar artery nourishes some of the brain's most important areas. Russell worried that clots and cutoffs of blood in the vertebral arteries would impair the work of the basilar artery and its downstream branches deep inside the brain.

The drop in blood flow was known to produce a variety of strokes. Symptoms might include coma, eye problems, vomiting, breathing trouble, arm and leg weakness, and sudden falls—but by definition had little to do with language and conscious thinking. However, because strokes of the rear brain can severely damage the regulatory machinery that governs life basics, they can also result in collapse and death. Even so, the vast majority of patients survive the attack and go on to recover most functions. Unfortunately, in some cases, headaches can persist for years, along with such residual troubles as imbalance, dizziness, and difficulty in making fine movements.

The medical world of Russell's day worried about these kinds of strokes, including a prominent type that began in circumstances that seemed quite innocuous. At beauty salons, during shampooing, women at times would have their necks tipped too far back over the edge of a sink, reducing the flow of blood through the vertebral and basilar arteries. The risk was judged especially great among the elderly. With aging, the vertebral arteries can lose their elasticity and narrow, and the normally smooth neck bones can grow spurs. When the neck bends far backward, the bony spurs can compress or otherwise harm vessels already narrow and inelastic. In addition, the stagnant blood can turn into a small factory of clot production. When the neck returns to a more normal position and the flow of blood resumes, the clots can travel down the arteries, heading deeper into the brain before settling in a narrow vessel and blocking its flow. A small epidemic of strokes resulted in a diagnosis known as the beauty-parlor syndrome.

Russell warned of yoga dangers in the pages of the *British Medical Journal*, a mainstay of the field established in 1840, just as Paul was finishing medical school in Calcutta. He drew parallels between yoga and such recognized threats as the beauty-parlor syndrome, noting that some poses produce "extreme degrees of neck flexion and extension and rotation." He specifically cited the Shoulder Stand and the Cobra, displaying a good understanding of the field. In the Cobra, or Bhujangasana, "serpent" in Sanskrit, a student lies facedown and slowly rises off the floor, pushing the trunk upward with the arms and extending the head and spine backward. Iyengar, in *Light on Yoga*, suggests that the head should arch "as far back as possible." Photos show him doing just that, his head thrown back on a trajectory toward his buttocks—in other words, the kind of maneuver that Russell found worrisome.

Cobra, *Bhujangasana*

In the Shoulder Stand, the neck is bent in exactly the opposite direction, going far forward, with the chin deep in the chest, the trunk and head forming a right angle. "The body should be in one straight line," Iyengar emphasized, "perpendicular to the floor." Ever the enthusiast, he called the pose "one of the greatest boons conferred on humanity by our ancient sages."

Where Iyengar saw benefits, Russell saw danger. The postures, he said, "must for some people be hazardous." His choice of the word "must" betrayed the speculative nature of his worry—but one grounded in a lifetime of experience. Russell warned that the basilar artery syndrome could strike practitioners of yoga and went on to cite a shadowy complication—doctors might have a hard time discerning its origin. The cerebral damage, he wrote, "may be delayed perhaps to appear during the night following, and this delay of some hours distracts attention from the earlier precipitating factor, especially when there is a catastrophic stroke." In that case, of course, the deceased could give no account of prior activities.

His caution went to the inherent difficulty of understanding the cause of invisible brain injuries. We typically think of illness as focused on a particular body part—such as the heart or lungs. But the origins of strokes often lie relatively far away from where they hit, starting in the wilds of the bloodstream and ending in the brain. The gap, moreover, could involve not only distance but time—hours and sometimes days— as a clot worked its way downstream or as a damaged artery slowly became swollen and gradually reduced the flow of blood. Such complicating factors meant that, for a large percentage of strokes, physicians could discover no obvious explanation. Their medical term for such injuries was *cryptogenic*, meaning their origin remained a mystery.

That kind of uncertainty had long obscured the cause and the extent of the beauty-parlor syndrome. In essence, Russell was now asking if the same thing was happening with yoga.

His alert proved timely. Perhaps he was simply ahead of his day, or perhaps his warning opened the eyes of colleagues, or perhaps the growth of yoga was resulting in more injuries. For whatever reason or reasons, an American physician in the following year, 1973, made public a gruesome case study. The author was Willibald Nagler. He worked on Manhattan's

Upper East Side at the Weill Medical College of Cornell University. A world authority on spinal rehabilitation, he had counted President Kennedy among his patients.

In his report, Nagler described how a woman of twenty-eight, "a Yoga enthusiast" as he called her in the sketchy anonymity of clinical reports, had suffered a stroke while doing a position known in gymnastics as the Bridge and in yoga as the Wheel or Upward Bow (in Sanskrit Urdhva Dhanurasana). The posture begins with the practitioner lying on his or her back and then pushing up, balancing on the hands and feet and lifting the body into a semicircular arc. An intermediate stage can involve raising the trunk and resting the crown of the head on the floor.

Wheel or Upward Bow, *Urdhva Dhanurasana*

Nagler reported that the woman entered her crisis while balanced on her head, her neck bent far backward. While so extended, she "suddenly felt a severe throbbing headache," he reported. She had difficulty getting up. After she was helped into a standing position, she was unable to walk without assistance.

The woman was rushed to the hospital and found to be experiencing a number of physical disorders. She could feel no sensations on the right side of her body. Her left arm and leg wavered. Her eyes kept glancing involuntarily to the left. And the left side of her face showed a contracted pupil, a drooping upper eyelid, and a rising lower lid—a cluster of symptoms known as Horner's syndrome. Nagler reported that the woman also had a tendency to fall to the left.

Diagnostic inquiry showed that her left vertebral artery had narrowed considerably between cervical vertebrae C1 and C2, revealing the probable site of the blockage that resulted in the stroke. It also showed that the arteries feeding her cerebellum (the structure of the rear brain that coordinates the muscles and balance) had undergone severe displacement, hinting at trouble within. Given the day's lack of advanced imaging technologies, an exploratory operation was deemed necessary to better evaluate the woman's injuries and prospects for recovery.

The surgeons who opened her skull found that the left hemisphere of her cerebellum had suffered a major failure of blood supply that resulted in much dead tissue. They also found the site seeped in secondary hemorrhages, or bleeding. In response, the physicians put the woman on an extensive program of rehabilitation. Two years later, she was able to walk, Nagler reported, "with broad-based gait." But her left arm continued to wander and her left eye continued to show Horner's syndrome.

Nagler concluded that such injuries appeared to be rare but served as a warning about the hazards of "forceful hyperextension of the neck." He urged health professionals to show caution in recommending such difficult postures to individuals of middle age.

The next case came to light in 1977. The man of twenty-five had been in excellent health and doing yoga every morning for a year and a half. His routine included spinal twists in which he rotated his head far to the left and far to the right. Then, according to a team in Chicago at the Northwestern University Medical School, he would do a Shoulder Stand with his neck "maximally flexed against the bare floor," echoing Iyengar's call for perpendicularity in *Light on Yoga*. The team said the young man usually remained in the inversion for about five minutes.

One morning upon finishing this routine, he suddenly felt a sensation of pins and needles on the left side of his face. Fifteen minutes later, he felt dizzy and his vision blurred. Soon, he was unable to walk without assistance and had trouble controlling the left side of his body. The man also found it difficult to swallow. He was rushed to the hospital.

Steven H. Hanus was a medical student at Northwestern who became fascinated by the case. He took the lead and worked with the chairman of the department of neurology to elucidate the exact cause of the disabilities, publishing a study with two colleagues when he was a resident. The doctors saw many indications of stroke and, in their report, noted the similarity of the man's symptoms to those of Nagler's female patient. The man could feel little sensation on the right side of his body. His eyeballs twitched. His left arm and leg were weak, had poor coordination, and showed a prominent tremor when he tried to reach for something or move his hand or foot to a precise location.

During the physical examination, the doctors noticed on the man's back a series of bruises. The bluish discolorations ran down his lower neck across the C5, C6, and C7 vertebrae. Apparently, the team wrote in the *Archives of Neurology*, "these resulted from repeated contact with the hard floor surface on which he did yoga exercises." The bruises, the doctors added, were a sign of neck trauma.

Hanus focused on assessing the inner damage. Diagnostic tests revealed blockages of the left vertebral artery between the C2 and C3 vertebrae. The team found that the blood vessel there had suffered "total or nearly complete occlusion."

During the man's first week in the hospital, the left side of his face developed Horner's syndrome—the constricted pupil and drooping eyelid. Slowly, he regained his ability to walk, though his gait remained clumsy. Two months after his attack, and after much physical therapy, the man was able to walk with a cane. But the team reported that he "continued to have pronounced difficulty in performing fine movements with his left hand."

Hanus and his team concluded that the young man's situation was no anomaly or medical oddity but instead a new kind of danger. Healthy individuals can seriously damage their vertebral arteries, they warned, "by neck movements that exceed physiological tolerance." And yoga, they stressed, "should be considered as a possible precipitating event." In its report, the Northwestern team cited not only Nagler's account of his female patient but Russell's early warning. The concern was beginning to ripple through the world of medicine.

The next case showed its global spread. In Hong Kong, a woman of thirty-four practiced yoga faithfully. One day, shortly after doing a Headstand for five minutes, she developed a sharp pain in her neck and numbness in her right hand. A surgeon made an incorrect diagnosis and prescribed neck traction and physical therapy. Her symptoms got worse. The attacks of nausea and dizziness grew in severity. Eventually her troubles came to the attention of a medical team at the University of Hong Kong and Queen Mary Hospital.

By this point—some two months after the neck pain—the doctors found that the woman showed signs of disorientation and paralysis on the left side of her body, as well as an inability to feel sensations of touch. Her eyes displayed the jerky movements typical of a rear-brain stroke, and the physicians made that the provisional diagnosis.

The doctors repeatedly scanned the woman's brain with imaging devices over the next few days. But they found nothing, even as her consciousness began to ebb. Finally, the team located a region of tissue that appeared dead from lack of blood. It ranged over the pons, the thalamus, and the occipital lobe. The doctors sought to pinpoint the cause of the stroke by injecting dye into the woman's neck arteries and taking X-rays. The diagnostic images showed no problems in the vertebral arteries but a severe blockage in the basilar artery.

The doctors had put the woman on blood thinners and clot-dissolving drugs after the provisional diagnosis. Eventually she underwent intensive physical therapy as well. After a year, she regained strength on the left side of her body. But she still exhibited clumsiness in her left hand.

Jason K. Y. Fong, a young neurologist, led the analysis. In 1993, he and his colleagues reported that the woman's problems had probably begun when vertebral arteries in the C1–C2 region suffered a tear or a severe reduction in blood flow. That produced a clot, the doctors wrote in *Clinical and Experimental Neurology*, that eventually worked its way into the basilar artery and blocked the blood supply to her inner brain. They attributed the lack of visible damage in the vertebral artery to the likelihood that the exceptionally long period between the Headstand and the hospital admission "may have allowed sufficient time for spontaneous healing."

The delay in uncovering the woman's stroke and its likely cause bore lessons for the medical community, Fong and his colleagues argued. The main one was the importance of learning the inconspicuous details of case history, which if taken seriously could speed diagnosis and treatment. Their warning echoed Russell's observation about overlooking the origin of brain damage.

The gravity of the Hong Kong case, the team concluded, showed that yoga could pose extraordinary risks to human health. The doctors cautioned that postures in which the neck came under great strain could be "potentially dangerous or even lethal." The latter word is one that physicians, steeped in a culture of cautious optimism and dry understatement, tend to avoid if possible.

The spike in clinical reports made yoga strokes a common feature of medical concern. The danger was judged to be at least partly due to underlying weaknesses in the vertebral arteries of some individuals. But it was difficult if not impossible to know who was at risk. So the warnings spread. They appeared not only in medical journals but in textbooks as health specialists gained new appreciation of the threat.

Science of Flexibility, whose first edition appeared in 1996, featured a section called X-Rated Exercises. It linked strokes to poses that stretched the neck far backward, including the Wheel and the Cobra. In summarizing the medical findings, the book's author called the value of the postures too small "to justify the potential, although rare, risk of vertebral artery occlusion." He suggested avoidance.

Injuries due to yoga turned out to range far beyond nerve damage and strokes. Waves of practitioners were showing up in emergency rooms. The Consumer Product Safety Commission, in monitoring the hazards of modern life, runs a little-known detective service known as the National Electronic Injury Surveillance System. It samples hospital records in the United States and its territories. By 2002, its surveys showed that the number of admissions related to yoga, after years of slow increases, had begun to soar. The number of admissions went from thirteen in 2000 to twenty in 2001. Then, in 2002, they more than doubled to forty-six. By definition, all these episodes involved men and women (and in some cases children) who had hurt themselves badly enough to seek out emergency assistance.

The spike represented the tip of a very large iceberg, since the system of federal monitoring produced only a statistical sketch. Most emergency rooms lay beyond its reach. Moreover, only a fraction of the injured visited hospital emergency rooms in the first place. Many—perhaps most—went to family doctors, chiropractors, neighborhood clinics, drugstores, and various kinds of therapists. Some probably decided to avoid treatment altogether and deal with the injury on their own. Thus, many hundreds or even thousands of yoga injuries in the United States went unreported.

The 2002 survey, like that of any year, gave a brief description of each person and each injury. An analysis of the information on the forty-six patients showed that they ranged in age from fifteen to seventy-five years, with the average age being thirty-six. The vast majority—83 percent—were women. The main type of injury centered on the complicated amalgams of bone, tendon, and cartilage known as joints, including the wrist (mentioned six times), the ankle and foot (five times), the knee (five times), the shoulder (four times), and the neck (four times). The injury write-ups contained an area for brief comments, which tended to describe everyday pains, strains, and sprains. But the comments also disclosed a number of serious traumas. Six of the injuries involved dislocations and fractures.

The survey listed no strokes—their diagnosis would typically require detailed examinations that went beyond the simple capabilities of most emergency rooms—but in several cases listed symptoms that might have coincided with the precipitating damage. "Acute neck pain," read one write-up. "Collapsed to floor while performing yoga," read another.

The brief comments tended toward the kind of pithy diagnoses and observations heard in emergency rooms: "dislocated right knee," "hurt shoulder," "low back pains." The reports usually cited yoga in general as the cause of the accident but on occasion named specific poses. "Sharp pain in abdomen since doing Cobra," read one report. Another said a male patient fainted while doing yoga in a warm room, falling and hitting his head hard enough to produce a bruise.

The wave—whatever its true dimensions—represented a clear rebuke to the "mother's milk" argument. Facts can be stubborn things, and they now suggested that yoga had long involved not only celebrated benefits but a number of hidden dangers.

For most of the twentieth century, yoga in the West enjoyed news coverage that can be described conservatively as excellent. The discipline was portrayed as nearly miraculous in terms of promoting health. An analysis of American reporting in the *Columbia Journalism Review* found much of it fawning. For gurus and publishers, the favorable coverage was, as the *Columbia* analysis put it, "the stuff of dreams."

The year 2002 marked a radical shift in the tenor of the reporting as the surge in documented injuries stirred public discussion on the issue of yoga safety. The seeming oxymoron of yoga damage had reached a critical mass in terms of size and social resonance that now made the issue impossible to ignore.

Stories appeared on radio and television as well as in magazines and newspapers, including *The New York Times* and *The Washington Post*. The rising public debate and the accompanying journalism meant that the injured were no longer portrayed exclusively as the anonymous stick figures of medical reports and federal surveys but began to take on the colorations of real life.

Holly Millea, for instance, was a freelance writer living in New York City who prided herself on staying in shape. The petite runner of forty-one practically never got sick. *Body & Soul* magazine recounted what happened when, in 2001, she took up Ashtanga yoga. By August of 2002, Millea began to feel numbness and tingling down her left arm and into her first three fingers. The pain grew and hampered her ability to sleep on her left side. The magazine said that, at one point, she thought the problem might be her heart or even multiple sclerosis—which ran in her family. Finally, after one emergency room visit and two rounds of medical imaging, Millea got the diagnosis: One of her vertebral disks had begun to bulge, squeezing a critical nerve. The magazine reported that her doctor wanted to surgically remove the disk and fuse two vertebrae together if the numbness failed to go away on its own.

"I am sure this is yoga-related," Millea said in the article, which appeared in 2003 amid her trouble. "It's at the base of my neck, and I was doing Shoulder Stand a lot. I was doing it wrong, and I was pushing myself too hard." She blamed herself and her competitive edge rather than yoga and its physical demands. "I am a super-athlete, and thought I could do anything," she added. "But I took it too quickly. I still needed to take baby steps."

A number of stories came down hard on Choudhury and his hot yoga. An article in *The New York Times* said health professionals found that the penetrating heat could raise the risk of overstretching, muscle damage, and torn cartilage. One specialist noted that ligaments—the tough bands of fiber that connect bones or cartilage at a joint—failed to regain their shape once stretched and that loose joints could promote injury. Another said the mirrored walls of Bikram studios encouraged students to neglect the traditional inner focus of yoga for outer distractions and the pressures of a room full of competitive individuals, also courting injury.

Not long thereafter, Choudhury came out with his book *Bikram Yoga*. It said nothing about dislocations or nerve damage, despite the medical warnings and bad press. It also managed to ignore the accusations of his critics. The few references that Choudhury made to the topic of physical damage centered on how hot yoga worked quite beautifully to promote a safe experience. The heat, he declared, lets students "twist and stretch with less chance of injury."

The period around 2002 also marked a turning point in that some elements of the yoga community started to move beyond denial and evasion to address the issue of damage. To a degree, the bad publicity left few alternatives. Now, for the first time, a number of experienced yogis and yoga publications engaged in serious debate on how to handle the quiet epidemic and come up with safety guidelines. It marked a period of public introspection—with notable exceptions.

The famous gurus, for the most part, remained silent. Publicly, at least, it seemed like their objective was to avoid involvement in any particulars that could prove distracting, embarrassing, and possibly litigious. The candor tended to come from the community's lower ranks.

A leading forum was *Yoga Journal*. It ran a number of articles, including one in 2003 in which a teacher revealed her own struggles. Carol Krucoff—a yoga instructor, author, and therapist at Duke University in North Carolina—told of being filmed one day for national television. Under bright lights, urged to do more, she lifted one foot, grabbed her big toe, and stretched her leg into Utthita Padangusthasana, the Extended Hand to Big Toe pose. As her leg straightened, she felt a sickening *pop* in her hamstring.

The next day, she could barely walk. Krucoff found that she needed rest, physical therapy, and a year of recuperation before she could fully extend her leg again. "I am grateful to have recovered completely," she wrote in *Yoga Journal*, adding that she considered the experience "a small price to pay for the invaluable lessons learned." These included the importance of warming up and never showing off.

Extended Hand to Big Toe, *Utthita Padangusthasana*

One glossy feature in *Yoga Journal* devoted ten pages of colorful photos and prickly text to the risks. "Yogi beware: Hidden dangers can lurk within even the most familiar pose," read the headline. Judith Lasater, a physical therapist and president of the California Yoga Teachers Association, argued that most poses hold subtle menace. The inherent risks can become quite palpable, she wrote, "because you may not have the necessary knowledge, flexibility, strength, and subtle awareness to proceed safely."

On another occasion, Kaitlin Quistgaard, the editor of *Yoga Journal*, told of how she had reinjured a torn rotator cuff in a yoga class, her pain becoming a cruel presence for months. "I've experienced how yoga can heal," she wrote. "But I've also experienced how yoga can hurt—and I've heard the same from plenty of other yogis."

And, no doubt, from plenty of lawyers. In fine print, the magazine began to run a legal proviso: "The creators, producers, participants, and distributors of *Yoga Journal* disclaim any liability for loss or injury in connection with the exercises shown or the instruction and advice expressed herein."

• • •

To its credit, the magazine paid attention to strokes, although it did so somewhat defensively and superficially. "Proceed with Caution," read the headline of its one-page article. The large color photo showed students upended in Headstands. The neck of a woman in the foreground was backlit and stood out. The article said doctors had identified five risky poses: the Headstand, the Shoulder Stand, the Side Angle pose (which Krishnamacharya had hailed as a cure), the Triangle (which Iyengar had carefully aligned), and the Plow, or Halasana. It had students lie on their backs, lift their legs up over their heads and back down onto the floor, inverting the torso. The article said such poses were judged potentially dangerous because they "put extreme pressure on the neck" or resulted in "sudden neck movements."

Plow, *Halasana*

It said nothing of two other poses that physicians had identified as serious threats to the brain: the Cobra and the Wheel, both considered X-rated postures.

The article warned that yoga practitioners could mistake injuries of the vertebral arteries for simple migraines or muscle tension. The symptoms of deeper trouble, it said, included piercing neck pain, pounding one-sided headaches, and facial paralysis. "Warning signs," it cautioned, "can intensify for hours or even days before a stroke hits."

So far, so good. But the magazine then proceeded to downplay the threat by failing to put the issue in perspective. It said doctors had found injuries to the vertebral arteries from all causes (such as yoga, beauty salons, and chiropractors) to be rare—annually, a person and a half out of every hundred thousand.

This was accurate. But it ignored the big picture. If twenty million people in the United States did yoga—a standard figure—and if yogis suffered the injury at the same rate as the general population (a very cautious assumption, given all the neck twisting and bending), that meant three hundred yogis in the United States faced the threat of stroke each year, or three thousand over a decade. The magazine not only neglected that baseline figure but sought to put its readers at ease by stating that yoga was "the culprit in a minuscule number of cases."

The reassurance was empty because the medical world had exactly zero evidence about the frequency of such damage. In fact, no scientist had ever published a study on how often yogis injured their vertebral arteries. The question was far too esoteric to have received the kind of major funding that would be required to address a deep riddle of epidemiology. So the exact size of the problem with yogis in the United States was simply unknown. What was easy to estimate was its minimal extent—roughly three hundred yogis a year.

Seeking to further brighten the grim subject, the magazine said "treatment is simple" and called rates of recovery high. But that rosy prognosis required that it ignore the agonizing months and years of therapy, the hand tremors and the clumsy gaits, the patients whose arms continued to waver and eyelids continued to droop.

Then, in what was apparently meant to be more good news, it added: "Death results in less than 5 percent of the cases."

Here again, the figure was correct but misleading because it failed to put the number in perspective. If three hundred yogis in the United States suffered injuries of the vertebral arteries each year (the lowball estimate), 5 percent of that would be fifteen—fifteen yogis who lay dead after wounds to their vertebral arteries resulted in brain injuries serious enough to kill. And the real number of fatalities, despite the percentage being "less than" five, was probably higher given the large number of poses in yoga that involve extreme contortions of the neck. Maybe it was thirty fatalities annually, and maybe three hundred over a decade. Globally, the fatalities might number in the thousands. It was an open question.

The article ended with a list of cautions—listen to your body, move slowly, avoid thrusting or jerky motions, go up to the point of resistance but never beyond. Its last warning focused on the neck. It advised students, especially beginners, to avoid putting the relatively thin, upper part of the spine in a position where it had to support a lot of body weight.

The magazine's attempt to deal with the sensitive topic appears to have made few waves in the world of yoga practitioners. Outside of *Yoga Journal*, the article got no general notice on the Internet from blogs, studios, or magazines, unlike the magazine's aerobics news. It rapidly sank into the void of cultural forgetfulness.

The subject of stroke nonetheless proved to be a topic of continuing worry among yogis—even if the discussions were superficial. More than three decades after Russell's warning, after the clinical reports, after the crystallization of medical concern, after the debut of the X-rated exercises, and after the threat summary in *Yoga Journal*, practitioners could still get lost in a cloud of uncertainty.

In 2002, the Internet buzzed with discussion about a woman of thirty-nine who did Power Yoga nearly every day and had suffered two strokes that threw her into the hospital. Her doctors, a friend reported, called her yoga routine the apparent cause and advised her to drop the practice. The woman did a beautiful Shoulder Stand, her friend reported in a discussion forum. But she wondered if the identified source of the trouble could possibly be accurate.

"Misinformed and misguided," one discussant said of the attending physicians. "Blaming yoga for a stroke is absurd."

During this period, yoga in America felt the sting of bureaucratic oversight for the first time as states began to regulate the training of teachers. They did so under the banner of consumer protection, the effort expanding in step with the new disclosures and the rising debate.

Regulators said licensing the schools would let states enforce basic standards and protect customers who typically spend thousands of dollars on training courses, as well as improving the quality of the experience for their students. "If you're going to start a school," said Patrick Sweeney, a Wisconsin licensing official, "you should play by a set of rules."

A disturbing new kind of injury came to light even as states began their regulatory effort. The case involved a woman of twenty-nine who was undergoing teacher training at Kripalu, the yoga emporium in the Berkshires. One night, she was practicing the rapid breathing method known as Kapalabhati Pranayama, or Shining Skull Breath—the form of Breath of Fire that Bikram students do as a grand finale. The next day the woman awoke with shortness of breath and pain in her left chest. Her symptoms slowly worsened, and she was taken to the Berkshire Medical Center, just up the road from Kripalu in Pittsfield, Massachusetts.

The doctors in the emergency room, upon seeing the woman's labored breathing and learning of her troubles, quickly put her on oxygen. The urgent question was what had gone wrong.

Lungs are like sponges that soak up air. They are highly elastic but largely passive. During a breath, the chest wall expands, forcing the sponge to draw in air. During exhalation, the sponge contracts and air goes out. It is mainly the action of the chest wall that governs the rhythms of the respiratory cycle. The sponge can do little on its own without the application of external force.

A quick X-ray showed that the woman was suffering from a serious failure of this mechanism, known as pneumothorax (from the Greek words for "air" and "chest"). The condition arises when air leaks into the space between the lung and the chest wall, loosening the usual grip of the wall and letting the sponge collapse. The lack of movement and breath can be life-threatening, especially if it involves both lungs. In the woman's case, the pneumothorax had partly collapsed her left lung.

In an emergency procedure, the doctors administered a local anesthetic, cut a hole between her ribs, and inserted a small tube that penetrated her chest wall and entered the pleural space. Then they extracted the unwanted air, allowing the chest wall to come back into play and her lung to reinflate. Immediately, the symptoms of labored breathing went away. The woman, after a week of recovery, underwent a procedure for removal of the tube.

In 2004, the doctors from the Berkshire Medical Center documented the unusual case in *Chest*, the respected journal of the American College of Chest Physicians. They noted that an imaging scan of the woman's chest had revealed no lung pathologies that might account for the pneumothorax, and concluded that the rupture was a direct result of yoga breathing. The case was without known precedent, they said, and showed that "adverse side effects can occur when one pushes the body to physiologic extremes."

In this case, the yoga community took notice and reacted. The days of denial and evasion were ending rapidly as the once-secretive topic of yoga injuries increasingly went public.

A yoga teacher and a medical doctor who had advised the teacher in developing a program for people with breathing disorders wrote a joint letter to *Chest*. The two, based in Sacramento, California, agreed that the rapid breathing exercise "most probably induced the pneumothorax" and backed the report's cautionary advice. But they said its warning about pushing the body to physiologic extremes created a false impression that "appears to unjustly blame all yoga techniques. This is not appropriate for a discipline that has generally been practiced safely for not hundreds, but thousands of years."

The yoga teacher—Vijai P. Sharma—took to the pages of the *International Journal of Yoga Therapy* to discuss the case and argue for the relative safety of Kapalabhati and other yoga breathing exercises. But his argument was heavy with caveats. He drew a distinction between fast and slow breathing, saying yoga's quick styles posed greater risk. Fast breathing, he wrote, "may reinforce or worsen preexisting structural or functional problems." Finally, Sharma enumerated a long list of safety guidelines and heightened risk factors (diabetes, chronic hypertension, persistent head pain) that made fast breathing seem like it was generally a risky venture.

Unless students exercise "out-of-the-ordinary patience and self-control," Sharma warned, "rapid breathing techniques such as Kapalabhati and Bhastrika are likely to be performed incorrectly and prove harmful in the long run."

Yet another case that came to light featured an aging yogi. The man had done yoga since his thirties and was sixty-three when the trouble hit.

His daily practice included the Headstand. He suffered no neck or back problems until one day he began to feel tingling and numbness in his fingers and toes. Over a few months, his legs and arms grew increasingly weak, and he began to experience the urge to urinate frequently.

His doctors saw the symptoms as classics of quadriplegia—limb weakness due to an injured spinal cord. Diagnostic imaging showed a region of disk compression and displacement between the C3 and C6 vertebrae. Health professionals, they wrote in their 2007 report, "need to be aware of this potentially serious complication of a relatively innocuous exercise."

In my thirties, I somehow managed to rupture a disk in my lower back. The cause seemed to be the repeated shocks of running on pavement rather than yoga. I looked into surgery, but found I could prevent bouts of pain with a selection of yoga postures and abdominal exercises.

In 2007, I experienced my own "serious complication" while studying with Robin in Pennsylvania. It happened as I did the Extended Side Angle pose, or Utthita Parsvakonasana. That was the posture that Krishnamacharya praised as a cure for many diseases. I was coming out of the pose and chatting with my partner—instead of paying attention to what I was doing—when my back gave way.

Blinding pain forced me to ignore everything but the explosion of fire. It was excruciating. My legs failed and the room vanished in tears. My body slammed into a wall.

Recovery took weeks. But the humbling experience gave me a deeper appreciation for yoga safety.

The redesign of poses by the yoga community ranged from tweaks to wholesale rearrangements. More drastic, some authorities called for the removal of risky postures from the Hatha corpus entirely or gave them warnings harsh enough to serve as de facto prohibitions, as with Kapalabhati and Bhastrika. The wave of new precautions was different from when medical outsiders drew up lists of X-Rated Exercises. It featured some of yoga's biggest names, giving it disciplinary cachet. Even Iyengar got involved. Moreover, the stars often made their recommendations in the literature of yoga rather than medicine, meaning the advice tended to receive wide readership among everyday practitioners.

The Headstand became an early target. In general, teachers advised students to unburden the neck. But they seldom mentioned that such easing contradicted Iyengar. "The whole weight of the body," the guru wrote in *Light on Yoga*, "should be borne on the head alone and not on the forearms and hands."

Richard Rosen—a teacher in Oakland, California, who had studied at the Iyengar Institute in San Francisco—called for exactly the reverse, with the complete elimination of weight from the head and neck. The idea was to suspend the head off the floor by pressing the forearms down. "If everything feels relatively comfortable," he wrote in *Yoga World*, "slowly lower your crown to the floor until it just barely touches. Keep 95 percent of your weight on the forearms and shoulders." His recommendations seemed to require a level of gymnastic skill and strength that many beginning and intermediate students would find daunting. As for the risks, Rosen never mentioned any specifically by name but simply called the Headstand "dangerous if not practiced intelligently."

Robin had us do Headstands in which we transferred body weight from the neck and head to the arms. With practice, it was fairly easy to do. "At this point in the game," he remarked as we practiced the redesigned pose, "you want a maximum amount of weight on your arms—and a minimum on your head." Someone asked how much weight should be transferred. "Seventy-two point three percent," he replied, eliciting howls of laughter.

Timothy McCall, a physician who became the medical editor of *Yoga Journal*, advocated a more drastic approach. He called the Headstand too dangerous for general yoga classes unless a teacher had a proven ability to avoid trouble. His warning was based partly on his own injury. Through trial and error, he had found that doing the Headstand had led to a condition known as thoracic outlet syndrome, which arises from the compression of nerves passing from the neck into the arms. As a result, he felt unusual tingling in his right hand as well as sporadic numbness. McCall stopped doing the pose and his symptoms went away. Later, in recommending that general yoga classes avoid the Headstand, he noted how the inversion could produce other injuries, including degenerative arthritis of the cervical spine and retinal tears because the Headstand raises eye pressure. "Unfortunately," McCall concluded, "the negative effects of Headstand can be insidious."

Today, a number of schools avoid teaching the inversion or ban it outright. The cautious styles include Kripalu, Bikram, Viniyoga, and Kundalini. If, as Iyengar claims, the Headstand is "the king of all asanas," its kingdom has undergone much contraction.

Other postures that have suffered banishment in some circles include the Full Lotus—one of yoga's most venerable poses. "Knees are hinge joints, meant only to bend and straighten, not twist," Dawn MacLear, a yoga teacher in Washington, DC, told the readers of *Health* magazine.

One of the most prolific reformers is Roger Cole, an Iyengar teacher with degrees from Stanford and the University of California who specializes in yoga anatomy and safety. He writes extensively for *Yoga Journal* and has spoken on yoga safety to the American College of Sports Medicine. Notably, Cole has drawn consistently on science to document the risky aspects of yoga postures and recommend safe practices.

In one column, he discussed how to reduce neck bending in the Shoulder Stand by lifting the shoulders on a stack of folded blankets and letting the head fall below that level, as we had practiced in Robin's class. In theory, that could increase the angle between the head and the torso from 90 degrees to perhaps 110 degrees. Cole also voiced rare criticism of Iyengar. He said the guru in *Light on Yoga* may have "inadvertently contributed" to neck injuries by calling for a perfectly vertical Shoulder Stand. Instead, Cole wrote, teachers should instruct students "to rest their weight toward the back of their shoulders and jackknife the body enough to take pressure off the neck."

Cole ticked off the dangers of doing the Shoulder Stand without such precautions. His list included muscle strains, overstretched ligaments, and cervical disk injuries. Strangely, he said nothing about strokes.

Eventually yogis sought to map the world of injuries by means of practitioner surveys. The questionnaires promised a better overview than the statistical surveys of the American government and, in the hands of yoga professionals, a better foundation for posture refinement and reinvention. The investigators, as was the case with many reformers, usually had backgrounds that combined yoga and science.

In 2008, yoga researchers in Europe published a survey about practitioners of Ashtanga—the fluid style from Krishnamacharya that Holly Millea had practiced. Their study limited itself to damage of the muscles and skeleton, involved practitioners only in Finland, and produced just one hundred and ten responses. But the results were fairly dramatic. The majority of the responders—62 percent—said they had suffered at least one injury that had lasted longer than a month, and some reported multiple upsets. The injuries were mostly sprains and strains, as well as two dislocations.

In 2009, a New York City team based at the Columbia College of Physicians and Surgeons published a far more ambitious survey of yoga teachers, therapists, and doctors around the globe. It was done in cooperation with the International Association of Yoga Therapists, the Yoga Alliance, and Yoga Spirit, an education group in Toronto. More than 1,300 people in thirty-four countries responded. The Columbia survey asked not only for personal experiences but also observations about yoga students and patients. It appeared in the *International Journal of Yoga Therapy*, ensuring its wide reading among yogis interested in healing.

The participants reported practicing Hatha and its many offspring, including Vinyasa, Iyengar, Anusara, Ashtanga, and Kripalu. The survey's central question—"What are the most serious injuries (disabling and/or of long duration) that you have seen?"—produced a number of revelations.

The largest number of injuries (with 231 reports) centered on the lower back. In declining order of prevalence, the other main sites were the shoulder (219 incidents), the knee (174), and the neck (110).

Amid these generalities came more specific accounts. The respondents said they knew of forty-three times that spinal disks had herniated, seventeen times that bones had fractured, and five times that practitioners had suffered heart problems.

Then came stroke. Its debut in a yoga survey came nearly four decades after Russell's warning. The respondents said they had witnessed four cases—in other words, they knew of four occasions in which yoga's extreme bending and contortions had resulted in some degree of brain damage.

For the community, the admission was a significant step. Few yoga books ever spoke of the danger—or looked into the medical literature—and the grim topic seldom made the upbeat pages of yoga magazines. Now, a major survey done by yoga professionals had documented the threat. It was an honest first.

• • •

Another surprise centered on judgments about what explained the injuries. The choices for survey takers included such factors as large classes, too much effort, and expanded ranks of students. A vast majority of the respondents—68 percent—pointed to "inadequate teacher training." That was remarkable because most were teachers. In effect, they were criticizing themselves and their peers.

The candor went to an inconspicuous deficiency of modern yoga—that teacher training varies enormously in quantity and quality, from slapdash to rigorous. You can get certified as an instructor with as little as 100 hours of training and even do the course entirely online, putting in no time whatsoever in a classroom and getting no supervision from an experienced teacher.

Today many popular styles adhere to the minimum standards set by the Yoga Alliance, a private group in Arlington, Virginia, that seeks to build public confidence in yoga. Its definition of a yoga teacher is anyone who has participated in at least 200 hours of real training. Still, that effort—equal to four or five weeks—seems like an extraordinarily low bar in terms of serious education. Would you study with a violin teacher who had trained for a month? A sculptor? A basketball player?

Bikram is more demanding. It trains its instructors for nine weeks. Yoga Alliance also endorses a category of training that requires at least 500 hours—equal to about three months.

Compare that to Iyengar. It requires candidates for teacher training to have studied the style for a minimum of three years, and then trains them for a minimum of two years and administers two examinations to ensure the requisite progress. As we have seen, it is the Iyengar people who have redesigned some of yoga's most dangerous poses. Teacher training puts much emphasis on how to lessen the risks.

My reporting on yoga injury kept producing surprises, none bigger than those involving Glenn Black. In late 2009, almost a year after meeting him, I received an email. Black said he had undergone spinal surgery. "It was a success," he wrote. "Recovery is slow and painful. Call if you like."

I caught up with him at Plaza Athénée. He said the surgery had taken five hours, fusing together lumbar vertebrae three, four, and five. He would eventually be fine but was under surgeon's orders to take it easy with his lower back. His range of motion, he added, would never be quite the same.

He had done it to himself, Black insisted. The injury had nothing to do with trauma or aging but instead had its origins in four decades of extreme backbends and twists. To me, that recalled the wear and tear injuries of the elder yogi whose Headstands led to limb tingling, as well as Timothy McCall and his arm troubles.

In Black's case, it appeared that long practice had resulted in spinal stenosis—a serious condition in which the openings between vertebrae begin to narrow, compressing spinal nerves and causing severe pain. Black said he felt the tenderness start twenty years ago when he was coming out of such poses as the Plow and the Shoulder Stand. By 2007, the pain had become extreme. One surgeon said he would eventually be unable to walk. So the master teacher—a man who had been quick to speak of yoga's dangers and to put down instructors who underwent stealthy rounds of reconstructive surgery—prepared to go under the knife.

Black said he became quite moody as the date approached. "It was incredible," he said. "There was a lot of turmoil."

I asked if the problem couldn't have been congenital or the result of aging. No, he argued. Black said he was sure he had done it to himself.

After recovery? "I'm going to be a decrepit old man," he joked.

The moral?

"You have to set aside your ego and not become obsessive," he replied, instantly serious. "You have to get a different perspective to see if what you're doing is going to eventually be bad for you."

Black said he had recently taken that message to a conference at the Omega Institute, his feelings on the subject deepened because of the surgery. But his warnings—articulated with new vigor—seemed to fall on deaf ears. "I was a little more emphatic than usual," he said. "My message was that 'Asana is not a panacea or a cure-all. In fact, if you do it with ego or obsession, you'll end up causing problems.' A lot of people don't like to hear that."

Black said the results were entirely predictable. "More people are finding out that yoga is causing injuries."

V

HEALING

Loren Fishman kept getting seduced. He studied mathematics, logic, and philosophy at the University of Michigan and Oxford and, on the side, dabbled in yoga and meditation. But he kept getting attracted to new subjects, kept wanting to learn something that would help him minister to what he considered an ailing world. In England, he stumbled upon Iyengar's book and became entranced. The yoga was so clean, so anatomically sophisticated, so advanced compared to anything else he had encountered. The book became his bible. At the same time, Fishman kept up his wanderlust, going to India to learn Sanskrit and ancient grammar so he could look for clues to the origins of mathematics.

One day in 1973, a friend mentioned that Iyengar lived nearby. Ah, Fishman thought—the master. The young man from Chicago was already a long way from home. But he wanted to go farther.

Fishman knocked on Iyengar's door and was surprised when the yogi himself answered.

"What do you want?" Iyengar growled.

"I want to learn your yoga."

"Why?"

"Because I want to heal."

"That's my great thing! Come in."

Fishman told the yogi of his dreams and then checked into a nearby hotel. It had a lush rooftop garden and everything an itinerant scholar could possibly want. As a teacher, Iyengar was tough and prickly. He would tease Fishman, saying the eager student had deluded himself into thinking he understood the Iyengar system.

Finally, after a year of instruction, Iyengar ordered Fishman to go home and start teaching. Fishman did so. But he also decided to get as serious as he could about healing and went off to medical school—Rush

Presbyterian St. Luke's in Chicago, one of the oldest and most respected medical schools in America. It was founded in 1837, two years after the inauguration of the Bengal Medical College, a half a world away.

Today a growing number of physicians study yoga after medical school. Fishman was one of the first to do so beforehand. His medical immersion when he was already an advanced student of the discipline let him see yoga through a Western lens, understanding its routines in terms of the fine distinctions of anatomy and physiology, chemistry and physics. In some respects, he was able to fuse the West's storehouse of scientific knowledge with the inspiration of Iyengar's visceral creativity. It was a fresh approach that bristled with possibility and seemed to offer a new way to minister to the world.

Fishman graduated in 1979 and did a psychiatric internship at the Tufts Harvard Medical Center in Boston, eager to help troubled minds. But he found it unsatisfying and instead threw himself into rehabilitative medicine.

The field seeks to help individuals with broken bones, torn muscles, dead nerves, injured tissues, and other physical disabilities. Its rehabilitations draw on a wide variety of tools and treatments. Standard ones include slings, braces, crutches, drugs, prostheses, walkers, physical training, therapeutic exercise, and many adjustments to the lives of patients.

In Fishman's case, the options include yoga.

Few locations in the world of medical real estate are classier than 1009 Park Avenue in New York City, between Eighty-Fourth and Eighty-Fifth. It has a large but discreet awning and the usual glint of polished brass. The old building is as elegant as any on the Upper East Side. The office sits amid a constellation of physician specialists up and down the wide boulevard and across the street from the tall spire and Tiffany glass of the Park Avenue Christian Church, a neighborhood icon. The Plaza Athénée lies twenty blocks south.

Inside, nothing suggests the office is special. It has the usual art and magazines. The main hint of individuality lies behind the receptionist—a large shelf of Fishman's books on yoga therapy. They discuss how to treat everything from multiple sclerosis to sciatica, the condition in which irritation of the sciatic nerve causes pain to radiate through the buttocks and down the leg.

I learned of Fishman while looking for ways to strengthen my back. His book *Relief Is in the Stretch: End Back Pain Through Yoga* prescribed what seemed to be a sensible regime of postures. I especially liked his explanation for what stretching did. He told of a hidden interplay between two kinds of sense organs woven into the body's tendons and muscles. As a muscle stretched, he wrote, the two systems sent conflicting signals. Contraction was stronger than relaxation, so the muscle stayed tense. If the stretch continued, however, that signal began to diminish of its own accord and the relaxation impulse started to dominate. The transition took time, Fishman wrote. It started as the stretch continued from fifteen to thirty seconds, and the relaxation signal grew to dominate in less than two minutes.

That mechanism, Fishman wrote, is why students of yoga should hold poses patiently—at a minimum, fifteen seconds to two minutes. Only then can the muscle relax enough to stretch farther. He said the lengthening can dramatically help victims of back pain. It can increase the range of normal motion, relaxing the spinal regions and leaving them more supple, flexible, and resilient. And that in turn can help avoid conditions that lead to muscle spasm—the sudden, involuntary contraction of muscles, sometimes accompanied by great pain.

I enjoyed not only Fishman's clear writing but also his earthiness. Many yoga books use fashion models to illustrate the poses. Fishman, lithe and limber, often modeled them himself.

He appeared in the reception area, saw a patient, and gave her a hug and a few encouraging words. He was short and wiry, a bundle of energy, his smile quick. He wore a pale-blue shirt in a checkered pattern and a discretely colorful bow tie.

We went into his office and he spread out trays of carryout sushi. The wall behind his desk was cluttered with the usual diplomas as well as a large photograph of Iyengar. The famous yogi sat in a Full Lotus, his head high and eyes open, a picture of pride and vitality. Nearby photos also showed two grown children in graduation caps and gowns. From my research, I knew Fishman had achieved quite a bit in his career. In addition to treating patients, he held a clinical professorship at the Columbia College of Physicians and Surgeons, the medical arm of Columbia University. He had published more than a hundred papers and articles. At one point, Fishman led the New York Society of Physical Medicine and

Rehabilitation, serving as its president. His thriving practice, Manhattan Physical Medicine and Rehabilitation, employed physicians at four offices around the city—on Park Avenue, on the Upper West Side, in Queens, and on City Island. He also served as treasurer of the Manhattan Institute for Cancer Research, a charity.

While eating, Fishman told of his therapeutic work. Often, he would jump up from his desk to show what he meant, either doing a pose or demonstrating his point on a human skeleton hanging nearby. He said he was sixty-six but looked to be in his fifties.

Yes, he said, he learned much from Iyengar. But as Fishman spoke, it became clear that his guru was no guru in the sense of being a role model he followed slavishly. Instead, Fishman honored his mentor by exhibiting the same kind of pigheaded independence that Iyengar did, trying things on his own, experimenting on himself and his patients, arriving at cures and treatments in a roundabout way. It seemed that Iyengar provided the context, not the content. Fishman seemed to be a modern thinker who liked to tinker, a kind of Thomas Edison of yoga therapy.

He told of his own painful experience with torn rotator cuffs, and how that led to what he called a miracle cure. He used the phrase with a wry smile.

The shoulder is the most flexible joint of the hundred and fifty in the human body. It lets the arm achieve an astonishing range of motion—up, down, sideways, rotated—through a clever but risky stratagem that centers on a shallow ball-and-socket joint. The rounded head of the humerus, the main arm bone, rests in a very modest socket on the scapula, or shoulder blade, which children like to call angel wings. The shallowness of the socket gives the humerus wide freedom of movement but also raises the risk of the ball popping out. The job of holding it in place goes mainly to the rotator cuff. Its four or five muscles (the number depends on the authority) originate on the scapula and fasten to the head of the humerus through the tough cords known as tendons. Atop the humeral head, the tendons merge to form a taut cap of connective tissue that not only holds the shoulder tightly in place but also, in something of a contradiction, helps move the arm.

Tears of the rotator cuff usually involve the tendons, limit arm motion, and can be quite painful. Athletes who raise their arms in repetitive patterns—swimmers, tennis players, baseball pitchers—know the problem well. The tears most often occur in the tendon of a muscle known as the supraspinatus, which lies above the rotator group. Its name derives from its origin just above a bony spine that runs across the scapula.

Fishman said he tore his right rotator cuff while skiing. He had surgery on it and then, a few years later, tore his left rotator cuff as well. His surgeon judged the tear quite serious and suggested he get an appointment with the city's best specialists. It was a total supraspinatus tear, and Fishman experienced the usual pain and arm limitations. Without assistance, he could raise his arm no higher than eighty degrees—a bit less than perpendicular to his body.

One day at home, during the month-long wait for a surgical consultation (yes, even physicians get caught in that kind of delay), Fishman was doing yoga and decided to attempt a Headstand. He found he could do it. Getting his head down and arms into the right position was no problem.

"My wife said, 'What *are* you doing?'" he recalled. "I got up to tell her and found I could raise my arm. Before, I couldn't. I went to the office and did it again. It kept working."

Astonished, Fishman threw himself into a program of research and confirmation, including visits with top surgeons. Both said he no longer needed surgery and expressed bewilderment at how little science understood the mechanics of the arm.

At his own office, Fishman led an investigation into how the Headstand had achieved the cure. His main tool was the electromyograph—the heir to Jacobson's methods for tracking muscle activity. It let Fishman and his aides zero in on rotator activation. The team took measurements as he stood upright and on his head. The readings showed that two other rotator muscles had joined the action—the subscapularis and the rhomboid major. They engaged most when Fishman inverted his posture and proceeded to raise his shoulders—a key feature of the Iyengar Headstand. Iyengar taught that, once students were upside down, they should widen and raise their shoulders as far from the floor as possible. That extra lift turned out to be the main factor that produced the healing benefits.

Fishman concluded that the Headstand taught the other rotators to assume new roles. "It's training yourself to use a different muscle," he said, smiling, talking a mile a minute. By another name, it was muscle substitution—avoiding an existing problem by using other muscles.

Pleased with the results, Fishman decided to see if the benefits could extend to others. He asked the next patients with torn rotator cuffs if they would like to try the Headstand cure. Sure, ten answered. He and his assistants taught them an easy form of the inversion that they could do with the help of a folding chair.

Fishman's prescription? Do it once a day for thirty seconds. Nothing more. At the end of six weeks, he and his team checked the patients. To Fishman's delight, nine of the ten found they could move their arms like a person with a healthy shoulder. All decided to forgo surgery.

Sharon Williams, a development director at Dance Theater of Harlem, had come to Fishman with chronic pain in her right shoulder. It had ached for a month, and examination revealed a partly torn rotator cuff. After she started the Headstands, the pain went away and she found that her arm could once again move through its usual range of motion. It was a huge relief.

The results were surprising. Fishman and his aides published them so other health professionals could learn the trick.

I asked where else yoga could heal.

Fishman said it excelled in such things as osteoporosis—the disease of the bone that removes minerals and leads to increased risk of fracture. It often strikes older women and, without pain or symptoms or diagnosis, lies behind millions of fractures of the hip, spine, and wrist. Yoga stretching, he said, worked beautifully to stimulate the rebuilding of the bone. It happened at a molecular level. Stress on a bone prompted it to grow denser and stronger in the way that best counteracted the stress. Fishman said that for three years he had been conducting a study to find out which poses worked best to stimulate the rejuvenation.

"It's a big thing," he said of the disease. "Two hundred million women in the world have it and most can't afford the drugs," some of which produce serious side effects. By contrast, Fishman enthused, "Yoga is free" and completely natural.

"There are bad things in yoga," he volunteered. But not enough to outweigh the benefits.

Fishman knew the dark side in detail, it turned out. He told me about an injury survey that he and his colleagues were doing—the one based at the Columbia College of Physicians and Surgeons that documented hundreds of yoga injuries, including strokes.

On the plus side, Fishman said, yoga excelled at fighting the stiffness of arthritis. The inflammation and tight joints restrict movement, and yoga worked to increase the range of motion. As was his habit, Fishman had written a book about it, *Yoga for Arthritis*.

How often do you prescribe yoga for your patients?

Twice this morning, Fishman replied.

A woman in her late thirties had severe osteoporosis. The loss of bone mineral had weakened her frame and she had broken her foot four times doing exercises. Fishman prescribed a series of yoga stretches to be done flat on her back, lessening the chance of spinal fracture and providing a stimulus to help bring back the minerals.

Another woman, in her early forties, had severe neck pain. She also suffered from degeneration of the macula—the highly sensitive part of the retina responsible for central vision. Fishman suspected that her poor eyesight had caused her, a stock trader, to hold her head jutting forward and to the side all day long in an unnatural position that had resulted in her neck troubles. He prescribed yoga positions in which she would lie on her stomach and raise her head up and backward—the motion opposite of her daily grind. He said it would strengthen her neck and counteract the degeneration, letting the damaged tissues heal.

"I prescribe a lot," Fishman said. He had an unfair advantage over most yoga therapists, he added, because he could use all the diagnostic tools of modern medicine to pinpoint the problem, after which he could come up with yogic remedies of unusual specificity.

"A lot of yoga therapists don't have that ability," Fishman said. "They treat in a very generic fashion that can be dangerous."

The woman that morning with the neck pains illustrated the importance of good diagnostics, he said. An electromyogram revealed nerve damage in her neck and allowed him to prescribe the right physical treatment. By contrast, a diagnosis that was more informal might have blundered into a yoga treatment with false promise and possibly bad side effects.

Fishman said he never distributed handouts showing yoga postures, though sometimes he handed out yoga books. Instead, he said he gave out prescriptions of the kind used for drugs and medicines. But instead of writing down names of pills, he drew pictures.

He searched his desk, found a pad, and sketched away. After a minute, he showed the results. It was a three-step plan for battling spinal stenosis—the condition that had struck Glenn Black, where the spinal canal narrows, causing serious problems.

His sketch featured happy little stick men. The first stood, arms out, and the next frame showed it bent over sideways, hand to foot. The second sat upright on the floor with one leg stretched out and an arm reaching back in a spinal twist. The third lay flat and lifted the legs with a belt. The stick figures were informative but only rough outlines. Fishman said his usual method was to go over the details with patients.

Every Tuesday in the late afternoon, he held a yoga session at his Upper West Side office. He called it a three-ring circus and invited me to visit.

The office was slightly chaotic amid the transition from regular hours to yoga therapeutics. Patients came and went. A portly man hobbled around on crutches, his leg in a large cast. A young man sat on the floor, rubbing a bad ankle. "Undefeated in the Playoffs," read the back of his bright red T-shirt. A big cardboard box overflowed with colorful yoga mats. The receptionist folded up a room divider and the area suddenly became large enough for a small class.

Patients drifted in, put down mats, and began stretching. Maybe six or seven showed up, from their twenties to their sixties. There were also two yoga teachers, both women. One was a regular assistant. The other had recently met Fishman at a meeting in Los Angeles on yoga therapy and wanted to observe him in action.

Fishman came in, bouncy and engaging, immediately the ringmaster. He chatted and led warm-ups, wearing bright yellow gym shorts and a gray muscle shirt. Nothing about him appeared to be sixty-six.

When the visiting teacher volunteered that she had recently had surgery for a bunion—the painful curvature and swelling of the big toe—he showed us a simple treatment. It consisted of stretching both toes toward each other and then back to their normal straightforward positions; back and forth, back and forth, stretched and relaxed.

Everyone tried it. He said the exercise worked to strengthen a specific muscle, the abductor hallucis. On the sole of his own foot, he showed us its location and confidently predicted that pumping it for twenty to thirty seconds each day would prevent bunions and might reduce or undo them. Fishman said he developed the method four years ago after discovering a bunion forming on his own foot. It went away. He predicted that the yoga teacher would never need surgery on her other foot if she did the exercise. Fishman added that, for a study, he was tracking about twenty patients with bunions who regularly did the stretch. "It seems to be working," he remarked.

Fishman divided the class into groups. In the smallest, his assistant worked with a petite woman who had multiple sclerosis. This degenerative disease of the central nervous system leaves its victims weak, numb, poorly coordinated, and prone to vision, speech, and bladder problems. Fishman wrote a book on the disease with Eric L. Small, a Los Angeles yogi who at the time of their collaboration had fought multiple sclerosis for more than a half century and had long found relief in yoga. Their recommended routine had nothing to do with fostering cures and everything to do with promoting a better quality of life—trying to reduce handicap and disability, increase safety, lessen fatigue, strengthen muscles, increase range of motion and coordination, improve balance, raise confidence, and promote inner calm.

Teacher and patient began the session in a standing position. The workout was warm, informal, and quite different from the traditional rounds of yoga postures.

The yogini, Rama Nina Patella, had the patient start by holding on to the top of a file cabinet and bending down, stretching her arms and back in a fashion similar to what would happen in Downward Facing Dog. It didn't work. The left side of the patient's body was beginning to atrophy, and her left hand had a hard time gripping the file cabinet. So Patella had her try again. Only this time the patient held onto Patella's hips, and Patella clutched her arms. It worked. The patient was able to stretch down, long and slow. "Take your thighs back. Stretch this arm out as much as you can," Patella said of the weakened side. "Keep breathing. Reach with this arm, the arm that's kind of unwilling. Stretch that arm. That's good."

After a minute or two in that pose, the patient stood back up, beaming.

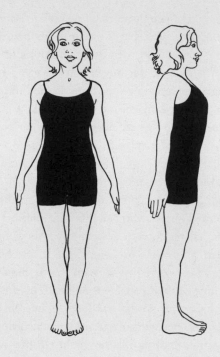

Mountain, *Tadasana*

Patella had her do the Mountain pose, or Tadasana. From the out-side, the pose seems simple and inconsequential. The student just stands there. But done right, it actually involves the subtle rearrangement and realignment of the whole body from head to heels, with muscles tensing and pulling and unbending the bones, the neck straight, the shoulders broad, the breath relaxed.

"Press your feet into the floor and lift your chest," Patella said. "You want a feeling in your feet like the roots of a tree growing into the earth and, from that rooted action, uplifting. Chest is open. Your shoulders are back. Let your breath flow as freely as possible. Good."

The patient had her eyes closed, concentrating, lifting and stretching. Her usual list to the left was somewhat diminished. She smiled.

Elsewhere, the room pulsed. A man in the class had, like me, herni-ated the disk that lies between the fourth and fifth lumbar vertebrae. Fishman had him doing a series of spinal extensions and elongations, and had the visiting yoga teacher shower him with attention.

As for himself, Fishman worked with a group of three women who, he said, had various kinds of abdominal problems. One suffered from prolapse—a condition where the uterus falls out of place, descending from the pelvis into the vagina. Normally, the muscles and ligaments of the pelvic floor hold the uterus in place. Uterine prolapse occurs when the muscles and ligaments weaken and stretch, undoing the usual support. Treatments include surgery, exercise, lifestyle changes, and a device worn inside the vagina that props up the uterus. Fishman took a direct approach that addressed the roots of the problem by seeking to strengthen core muscles and abdominal support.

He showed the women how to do a variation of the Warrior pose, or Virabhadrasana. From a standing position, he moved one foot forward and the other back, raised his arms straight up, and bent his forward knee. The result was the slow lowering of his pelvis as well as the stretching of his legs and abdomen. "Then you come down," Fishman said, dropping the thigh so low that it formed a right angle with the back. "Like this." He stretched his arms high up and his pelvis down low. Then the women tried.

Warrior, *Virabhadrasana*

Fishman moved among them, offering words of advice, encouragement, and—sparingly—praise. He exuded confidence and encouraged them to try hard. "Stretch up as high as you can," he urged, "stretching way up, way up. Good."

After a pause, Fishman led the women into another Warrior variant. It required not only stretching but balance. From the first pose, he had them stand on one leg while raising the rear leg to a horizontal position and lowering the arms and torso. It was like Superman flying with one leg extended straight down. Fishman moved among the women, offering alignment tips. "Bring this hip down," he told one woman, lightly touching the hip. She quickly rotated her hips into a horizontal plane.

"Good. With the hip down, raise the leg up." He put his hand under her leg, signaling how he wanted her to raise it, and she gave a little moan at the effort. "See what you're doing?" he asked. "You're stretching everything in here"—he motioned to her lower torso—"front and back."

And so it went. For the better part of an hour, Fishman led the women through numerous sitting and standing poses, all aimed at stretching and strengthening their midregions. "Try to engage those muscles," he said at one point, encouraging the women to push themselves even while paying attention to the sensations.

Fishman closed with a meditation. It began with a few minutes of relaxed breathing with eyes shut to foster inner awareness of body position and sensation, especially in the lungs.

"Feel on the right side and the left side," he said. "Is it the same? Feel your shirt against your skin. Is it pushing equally? How about the tenor of your breathing? Are you a soprano or an alto or a baritone? Listen to your breathing. Don't try to do anything. Just pay attention. How does the air go in? Both nostrils? One? Feel the bottom of your lungs, the sides, the back and front. Feel what's going on in there—these capricious things that we need so desperately and never see."

Then there was quiet.

Weeks later, I returned to Fishman's Upper East Side office to ask some follow-up questions. He said his staff was dismantling his West Side office for a bigger space around Columbus Circle. It would have a larger room for classes, Fishman said. The yoga aspect of his practice was clearly expanding.

He said none of the other doctors in his practice did yoga or prescribed it to patients. It was his specialty alone, though, he added, one of his aides and a physical therapist also studied the discipline.

I asked how, overall, yoga had aided his practice. He said it acted as a kind of laboratory for the nurturing of physical creativity, letting him experiment on his own body and that of willing patients to discover new kinds of natural cures and therapies. Without yoga, he said, "I'd lack the most interesting, least expensive, and most helpful and versatile form of treatment that I have."

I asked if he had ever had surgery on his left rotator cuff. No, he answered. The yoga solution, he added, had been working just fine for seven years now.

He held his left arm high over his head and smiled.

To become a physician, Fishman had to undergo an ordeal of schooling and formal assessment that, in the end, gave him admission to an elite club. The first big evaluation was the U.S. Medical Licensing Examination, a series of tests taken during medical school and residency. He then earned a medical license from the State of New York and its board of medical examiners. To stay in good standing, his license required that he do fifty hours of continuing education each year. He also earned professional certifications from the American Academy of Physical Medicine and Rehabilitation, the New York State Workers' Compensation Board, and such professional bodies as the National Multiple Sclerosis Society.

So, too, the physical therapists who work for Fishman are licensed through the state and the American Physical Therapy Association. These organizations require graduate degrees in physical therapy, as well as continuing education. At first the degree tended to be a master's, but the field of late has moved rapidly toward requiring a doctorate. The course work for such a diploma is heavy in embryology and histology, anatomy and physiology, pathology and pharmacology, kinesthesiology and imaging techniques. Many states require the dissection of cadavers.

The goal of mandatory licensure is to form groups whose members have met certain minimal requirements that—among other things—are meant to protect the public from harm. The highly regulated world of medicine is typically backed by the force of law and seeks to deter and penalize interlopers. In New York State, where I live, practicing medicine without a license is a felony punishable by up to four years in prison.

What Fishman underwent to become a yoga therapist bears no resemblance. He received no formal training, earned no license, faces no requirements for continuing education, and will never confront any oversight panel or threat of reprimand and penalization. His complete freedom of activity arises not because of any deficiency on his part but because the United States has no regulatory body for yoga therapy. None. Zip. Nada. Few countries do. The field is, on the whole, completely unlicensed and unregulated.

Even so, the yoga community has managed to foster the illusion that the United States has a system in place for the accreditation of yoga therapists. That New Age fiction is helping to promote the field's growth. Unfortunately, it is also deceiving people, some desperate for healing because of serious illnesses and injuries.

Yoga teachers who aspire to the role of contemporary healer often put after their names the initials RYT—short for Registered Yoga Therapist. They do it on books, brochures, and Internet sites. The practice may seem innocuously similar to how physicians use MD and dentists DDS. But the situation is entirely different.

Since the RYT terminology is unfamiliar to many people, yogis, yoginis, and yoga groups often spell it out. For instance, *Yoga Journal* regularly uses the phrase "registered yoga therapist" to describe its experts and authors. So, too, a Google search produces many hundreds of hits for the phrase, identifying local healers from coast to coast.

In 2006, the Montgomery County Department of Recreation, in the Washington, DC, suburbs, advertised classes featuring a registered yoga therapist "sensitive to individual needs." Authors and booksellers love the accolade. Publicity materials for *Yoga and the Wisdom of Menopause*, by Suza Francina, a popular yoga writer, call her "a registered yoga therapist with 30 years' experience."

And why not? The phrase sounds authoritative. The dictionary defines "therapist" as "a person trained in methods of treatment other than the use of drugs or surgery" and defines "registered" as "qualified formally or officially." A Registered Yoga Therapist would, presumably, have undergone extensive training and passed the rigorous examinations of a national body of health-care specialists.

Wrong. In fact, there is no such thing as a Registered Yoga Therapist. It is an illusion—perhaps in some cases a lie. The world of professional recognition *does* have a category that centers on the practice of registration, though it is considered the lowest step in the expert hierarchy—far less meaningful than, for instance, licensing by a state medical board. National groups that register professionals typically record only various types of personal information, such as name, address, and form of practice. So, too, applicants for registration usually face no requirements to establish their education credentials, to pass national exams, or to show other evidence of expert proficiency. Registration, in short, bears no comparison to the rigorous world of health-care certification.

Yoga therapists have adopted the evocative terminology through guile or sloppiness, or perhaps an unconscious mix of the two that plays to their economic self-interest. For whatever reason, many simply assert the status. In doing so, they can arguably draw cover from long discussions in the yoga community about the *possibility* of creating a regulatory category known as the Registered Yoga Therapist, as well as confusion over the meaning of similar credentials.

Yoga Alliance uses the acronym RYT as shorthand for Registered Yoga Teacher. Its listed teachers can legitimately use RYT after their name. But individual yogis and yoginis, in their advertisements and self-promotions, often morph the term *teacher* into *therapist*—a change the field's leaders actively discourage.

"A growing number of yoga instructors seem to be assuming the role of 'yoga therapist' without having had the necessary training and experience," Georg Feuerstein, editor of the *International Journal of Yoga Therapy*, conceded in a 2002 editorial. The term, he added, gets used liberally and often interchangeably with "yoga teacher." As a solution, yoga leaders around 2003 began discussing whether the alliance should expand its registry to include yoga therapists. Nearly a decade later, no such registry had materialized.

The International Association of Yoga Therapists, based in Prescott, Arizona, for several years has led public discussions of the possibility of creating standards as well as its own registry for yoga therapists. It has done so in the pages of the *International Journal of Yoga Therapy*, its publication. For instance, in 2004, John Kepner, the association's executive director, wrote an editorial arguing that registration with his group "should be a mark of high accomplishment, acceptable to those steeped in the Yoga tradition and credible to integrative health care providers." His article made repeated references to Registered Yoga Therapists.

But as of 2011, after more than a decade of discussion, nothing had come of the registration idea. Yoga therapy remains a free spirit. Anyone can claim to be a yoga therapist.

Individual schools have sought to fill this void (and their bank accounts) by teaching courses in healing and graduating what they call Certified Yoga Therapists. But the schools make up their own curriculum and teach whatever they deem appropriate, as does, for example, the Namaste Institute for Holistic Studies, in Rockport, Maine. *Namaste* is a Hindu greeting meaning "I bow to the divine in you." The school's program for certified therapists "provides in-depth training," runs for a month, and costs nearly four thousand dollars. As is the case with registration, no national body administers tests, awards certifications, polices the field, or sets rules for what constitutes minimal education requirements for Certified Yoga Therapists. Once again, anything goes.

"There is no such thing as a Registered Yoga Therapist," Kepner of the International Association told me. "And the schools offering certifications in yoga therapy provide widely different types and amounts of training—say, from eighty to eight hundred hours." Kepner also disparaged the registry idea as a "weak form of accreditation and credentialing, and not really sufficient to develop a credible professional field."

He said his group was investigating the conventional route to professional accreditation—as nutritionists, chiropractors, and acupuncturists have done successfully over the years. As a first step, he said, the association was supporting the formation of a council of schools that would establish a standard academic curriculum for the training of yoga therapists.

All that may sound quite reasonable and forward-looking. But the association has long engaged in activities that have helped blur the issue of what constitutes a genuine credential.

Every time members pay their annual dues to the association, they receive a fancy certificate suitable for framing that looks very much like a school diploma. It is personalized, too. I've gotten a number of them— one when I joined and others when I have renewed my membership (which now costs ninety dollars). The first one hangs on the door of my home office.

It looks quite elegant. The certificate is printed on parchment-colored paper and bears a gold border in a fine geometric pattern. The whole idea of a certificate—which the dictionary defines as "a document proving that the named individual has fulfilled the requirements of a particular field and may engage in its practice"—is evocative of professional accomplishment. The certificate's reference to an "award," and its twin signatures at the bottom, reinforce that idea. But a quick read shows that the document is in fact quite meaningless. In my case, it says I received the certificate "in recognition of supporting Yoga as an established and respected therapy in the West."

The certificate is signed at the bottom by Kepner, the group's director, and Veronica Zador, its president. I've seen similar ones displayed prominently in yoga studios—framed and lending an air of authority to the teaching and healing enterprise.

The phony credential does an injustice to the talented yoga therapists who have labored for years and decades to develop their healing expertise and have helped countless people. From what I saw, Nina Patella in Fishman's office ministered to a patient needing special attention with great skill and compassion. So did Amy Weintraub, the yogini who specializes in treating depression. The organizational ups and downs of the field reveal its troubled development but say little about the genuine therapeutic abilities of particular individuals.

Even so, the continuing lack of regulation and the hundreds of false claims that aspiring healers make about their credentials are helping fuel the field's rapid growth. The International Association of Yoga Therapists has seen its membership rolls increase from hundreds to thousands of members. Dozens of books hail yoga therapy as a sound treatment for most every kind of ailment—including cancer and AIDS, Alzheimer's and Parkinson's. And clients are lining up, ready to pay for what appears to be an all-natural, innovative kind of healing.

The surge is creating not only a lively commerce but, as the last chapter showed, a threat, since yoga in unskilled hands can bear the risk of serious injury. Patients can get hurt. In some cases, yoga therapists have managed to prescribe what turns out to be exactly the wrong move. As Fishman put it, "They treat in a very generic fashion that can be dangerous."

In 2008, *Yoga Journal* released a market study done by Harris Interactive. The survey of more than 5,000 people—a sample large enough to be considered statistically representative of the entire population of the United States—showed that yoga therapy had achieved wide acceptance among patients and, arguably more important, among the nation's health-care providers. The survey extrapolated to conclude that a doctor or therapist had recommended yoga to nearly fourteen million Americans—or more than 6 percent of the population. And nearly half of all adults reported that they held the field in high esteem, saying they felt yoga would help them if they were undergoing treatment for a medical condition. Even if the poll targeted yoga enthusiasts and overstated the degree of national interest, the trends nonetheless seemed quite real.

"Yoga as medicine represents the next great yoga wave," Kaitlin Quistgaard, the editor of *Yoga Journal*, stated during the study's unveiling. "In the next few years, we'll be seeing a lot more yoga in healthcare settings and more yoga recommended by the medical community."

Perhaps so. But for the moment, yoga therapists are wholly unregulated and thus the quality of their care is random. Some are geniuses. Some are charlatans. And many are surely mediocre and potentially dangerous, their heads filled with dreamy nonsense about healing and empty of real knowledge about the serious dangers of some poses. Yoga therapy is now in the Wild West stage of development. Some practitioners are busy putting up shingles, selling snake oil, and making astonishing claims. Buyer beware.

If the origins of the modern field can be traced to a single person, it would be Larry Payne, the founding president of the International Association of Yoga Therapists. Like Fishman, Payne came to yoga therapy early—decades before its current popularity. But his background is quite different from that of Fishman, and his long pursuit of professional credibility illustrates some of the difficulties that the field must overcome if its would-be healers are to become trusted members of the health-care community.

A native Californian of athletic build and interests, Payne began as an advertising executive in Los Angeles. It was a good life. Payne had lots of money and perks, including a generous expense account and a company car. By 1978, however, the rising pressures started to hurt. His blood pressure soared and his back went out.

The pain drove him crazy. (I can sympathize. I once got hauled away in an ambulance, blind with agony.) He tried orthopedic specialists, physical therapists, and drugs. Nothing worked. He looked into surgery. He felt like an old man, though only in his midthirties.

Desperate for relief, Payne let a friend drag him to a yoga class. He did the postures, the deep breathing, the relaxation. It was amazing. For the first time in two years, his back pain disappeared. He marveled at the unfamiliar feeling of happy relaxation. Overall, it was as if a huge weight had been lifted from his shoulders. In some ways, he felt reborn.

Payne continued the lessons and left his advertising job. Soon he decided to devote himself to yoga.

In India, he traveled to Madras (later known as Chennai) and studied at the Krishnamacharya Yoga Mandiram, a school of yoga therapy that had been recently founded by T. K. V. Desikachar, the son of Krishnamacharya, the guru to the gurus. The school, like its namesake, hailed yoga's therapeutic benefits and focused on healing. Among its specialties: the relief of lower back pain. It also treated everything from headaches and high blood pressure to asthma and schizophrenia. By 1980, Payne was hooked. He proceeded to recast himself with all the energy and marketing savvy of an ad executive.

In 1981, he founded a yoga center in Los Angeles that he named Samata, Sanskrit for "balance." It was located near Venice and Marina del Rey, two seaside playgrounds. Payne taught regular yoga. But he also toiled to advance the kind of healing that he himself had experienced and to integrate it into Western medicine. If nothing else, that was an astute business move that helped distinguish his enterprise from the region's growing number of yoga teachers.

The credential he needed for high credibility in his new calling was a medical degree. But the course work was staggering. The next best thing was a doctorate. It, too, could open doors. But either a doctor of philosophy degree or doctor of physical therapy degree represented a huge investment in time and money for a young person, much less a man of forty who was trying to reinvent himself. A solution beckoned. It was convenient, located just across the Santa Monica Freeway in Brentwood, home to the rich and famous. Payne found a book on alternative colleges that gave it a thumbs-up.

Pacific Western University had just one drawback. It was what federal investigators came to look upon as a diploma mill. The private school gave the appearance of being an institution of higher learning, but in reality provided little by way of education for its students. It accepted the transfer of academic credits and gave credit for life experience, but required no classroom study or instruction. What it did with enthusiasm was award master's and doctoral degrees—all for a flat fee. A doctorate cost a bit more than two thousand dollars. That was nothing compared to what a student could pay at a real school, semester after semester.

Pacific Western had no national accreditation, and that meant its degrees carried no weight with informed scholars and employers. In time, state, federal, and foreign governments came to regard the school as an educational fraud. Some states blacklisted its degrees as worthless or illegal.

In 1987, when Payne got his doctorate, the school was fairly new and had yet to receive much scrutiny. On his résumé, he declared that he "earned a master's degree and a doctorate in fitness education with an emphasis in Hatha Yoga from Pacific Western University." The school might have had no foreign language requirement or classroom challenges, but its message was relentlessly upbeat: "Look within yourself," the school told prospective students. "Exert yourself with dynamic will power through positive thinking and persistence and you will stretch your talents and imagination and achieve new heights of learning while attaining professional and personal success."

Payne did just that. He became a whirlwind. Armed with his new credentials, he gave lectures, wrote books, made instructional videos, appeared on radio and television shows, and, in 1989, helped found the International Association of Yoga Therapists. As its founding president, and later its director and chairman, he enjoyed a new world of global influence. The group began its journal in 1990, and Payne worked hard in his new job to drum up new readers and memberships.

Bigger things beckoned, and in 1999 Payne entered the burgeoning field of popular yoga books. *Yoga for Dummies* featured his Ph.D. prominently on the cover, along with that of his coauthor, Georg Feuerstein. As a team, Payne covered yoga's modern aspects while Feuerstein, an Indologist, handled its ancient ones. The twin credentials lent the book an air of authority and set it apart from competitors, although its biographical materials gave no indication of where Payne received his doctorate or in what field. The book identified him as chairman of the International Association of Yoga Therapists and a yoga teacher with a thriving practice who "responds to his clients' specific health challenges," implying that he was a credentialed healer. At a minimum, many readers probably assumed that his doctorate—ostensibly a proof of high academic standing—meant that the book reflected the best understanding of modern science.

The truth lay elsewhere.

The chapter on yoga breathing distinguished itself for its repeated praise of supplemental oxygen as a secret of yoga's powers. Deep inhalation, it declared, "loads your blood with oxygen." Three pages later, *Yoga for Dummies* enlarged on the error. Pranayama, it said, "allows you to take in more oxygen food for the 50 trillion cells in your body." That, of course, not only described a false oxygen rise but made the claim sound more authentic by linking it to the striking body-cell figure.

Two pages later, the book goofed again. After reminding the reader that yogic breathing "brings more oxygen into your system," *Dummies* raised a red flag. "Don't be surprised," it warned, "if you feel a little light-headed or even dizzy." That explanation, of course, went to the repercussions not of adding oxygen but of blowing off carbon dioxide, which can result in blackouts. It was another missed opportunity for understanding.

The trend culminated with the description of a breathing exercise that, the book assured, would "treat your body with oodles of oxygen." All the body's cells, it emphasized, "will be humming with energy and your brain will be very grateful to you for the extra boost."

Payne's book also managed to misrepresent one of the most fundamental ways in which yoga affects the human body. As we have seen, scientific investigators, starting in the nineteenth century, established that a defining characteristic of yoga—perhaps *the* defining characteristic—is how it can *slow* the body, the mind, and the overall metabolism to foster tranquility. Paul focused on hibernation, Behanan on the "retardation of mental functions," Bagchi and his colleagues on the "extreme slowing" of respiration and heart rates, Bera and his colleagues at Gune's ashram on lowered metabolism, many scientists on the body's parasympathetic brake, and Benson on wide physiological drops that led to hypometabolism. Yes, a few breathing styles—such as Bhastrika and Kapalabhati—can excite. But overall, they are the exception, not the rule. As the team of Indian scientists in Bangalore reported, the regular practice of yoga causes the resting metabolic rate to fall.

This physiological fact of life has an obvious social proof. Yoga has won a global following not because of some ostensible ability to zip people up but by its demonstrated power to slow them down. It has proved extraordinarily effective at undoing urban stress and the tensions of modern life. The reason yoga studios are so ubiquitous in big cities is because they are a great antidote to big cities.

Payne, citing no evidence, declared that the physical truth lay precisely in the opposite direction. Yoga breathing, he stated, "steps up your metabolism." He felt so confident about the claim that later in the book he generalized the energizing effect to the discipline as a whole. Practicing yoga, he asserted, will "boost your metabolism" and "helps you step up a flagging metabolism."

His misinformation helped pave the way for credulous authors to come, including Tara Stiles, the former model who authored *Slim Calm Sexy Yoga*. He gave the myth new energy.

Dummies took the muddled thinking about physiology and, like Stiles, applied it to a sensitive issue of personal appearance. The metabolic rise, Payne assured his readers, could aid their realization of one of the obsessive goals of modern life—maintaining a slim figure. The heightened metabolic state, he declared, was "the best manager of weight increase." His claim was remarkable. By implication, the word "best" put yoga above dieting, exercise, walking, general fitness, and wise nutrition as a means of burning calories and controlling weight. And, lest readers forget, he reiterated the slimming claim. Yoga postures, he said, "keep the rolls off your midriff." The discipline, *Dummies* stressed, "helps you shed surplus pounds."

That said, the pseudoscience of Payne on oxygen and metabolism comprised only a small part of his book. Most of his advice lay in standard postures and tips, anecdotes and encouragement. Photo after photo showed him—athletic and good-looking—going through the poses and helping students. Interestingly, he devoted the majority of the book to what he called "Health Maintenance and Restoration" but made few direct claims for healing.

It sold. Starting in 1999, *Yoga for Dummies* went through at least fourteen printings—far more than most yoga books. It became a standard reference for beginners. And, almost magically over the years, the point size of the font on the cover that announced Payne's Ph.D. grew larger.

In 2000, he traveled to Davos, Switzerland, and the World Economic Forum. He was, as a Samata news release put it, the first yoga teacher to address the group—a gathering of more than two thousand world leaders. The notables included Bill Clinton and Tony Blair, Bill Gates and the novelist Umberto Eco. Sunday morning at Davos is usually reserved for relaxation and sports. Many participants ski the nearby slopes. But not that Sunday—not with its driving snow, high winds, and zero visibility. Payne found his session packed.

He had arrived.

Around this time, Payne met a medical doctor with whom he formed a close relationship. The doctor, Richard Usatine, had practiced at the Venice Family Clinic and gone on to help run the program in family medicine at the University of California at Los Angeles School of Medicine—one of the world's top medical schools, located just a few miles from Payne's yoga center. The two met after Usatine walked away from a car accident but suffered serious back pain. He tried the usual treatments but got no relief. Soon he was referred to Payne and began a yoga routine that quickly ended his anguish.

The men bonded over backs, healing, and a deep belief in the body's hidden powers of recuperation. The two began discussing how to give medical students a sense of yoga's benefits and soon founded an elective course. A medical-school first in the United States, the UCLA class gave an overview of yoga and yoga therapy. The popular course became a regular part of the school's elective curriculum.

Encouraged, Payne and Usatine joined forces to author *Yoga Rx*, published in 2002. It taught how yoga could treat everything from heartburn to asthma to back pain. In bold type, its cover featured not only Usatine's M.D. but Payne's Ph.D. The former ad executive was moving not only ahead but up.

Significantly, *Yoga Rx* was more closely aligned to the science than *Yoga for Dummies*. The book, as it proclaimed from the start, reflected Usatine's medical expertise "on every page." In particular, it made none of Payne's false claims about pumping up the metabolism and burning more calories as a method of weight control, even though it devoted a long section to fighting obesity. The account was plain but honest.

As Payne moved ahead on his blur of projects, he managed to stay more closely aligned with the science. His encounter with a knowledgeable coauthor seemed to produce something of a midcourse correction.

His new moderation showed in a spinoff. In 2005, he released *Larry Payne's Yoga Rx Therapy: Weight Management for People with Curves*, a video disc. The program had little depth. And it made the customary omission of saying nothing about how yoga tended to lower the metabolic rate and, all else being equal, threatened to saddle the student with added pounds. But Payne, looking sincere, speaking with ease and confidence, made no wild claims and statements at odds with the known science. Instead, he voiced simple truths and encouragement.

"Real weight management is about making sensible lifestyle changes, including exercise," Payne said. He added that regular yoga built self-discipline. "You wouldn't think that doing yoga would keep you from opening the refrigerator door. But it does." He smiled.

Unlikely as it seems, the professional lives of Fishman and Payne have intersected and drawn close over the years. It is, one might argue, a kind of healing.

Fishman joined the Advisory Council of the International Association of Yoga Therapists, where he works with Payne in an effort to improve the profession's standards as well as its methods of practice and teaching. They go to conferences together, chat, and socialize.

Fishman told me he liked Payne. "He's good, knowledgeable, and serious—and a nice guy on top of it," Fishman said.

So, too, Payne kept up his science trajectory. In 2005, he helped found and became the director of the first program in the United States for the certification of yoga therapists at a university—Loyola Marymount, in Los Angeles, a Catholic school that overlooks the Pacific. The program has turned out dozens of therapists—mostly women—and sits on the Council of Schools that the association established as a way to promote standardized training.

Payne told me that his recent emphasis on science grew out of the community's rising interest over the years as well as his own. As for his Ph.D., he defended the degree as a substantive credential that he used in good faith and said he was unaware of anything improper about the school. "I honestly knew nothing about any shady stuff."

It seems clear that, early on, Payne was as much a victim of pseudoscience as a perpetrator. He did not create the blur of misinformation but simply immersed himself in it and proceeded to send it toward a large audience. He was credulous rather than duplicitous. That is not to say his missteps were inevitable. Iyengar and some other famous yogis managed to avoid the fog. But Payne failed to do so and became one of its prominent casualties.

Fishman knows all about hazards that lie beyond the strictures of modern medicine. He says he works with Payne, the association, and yoga therapists out of a desire to help them become more scientific.

"Yoga is in danger," he told me in his Manhattan office. "It can tip either way—toward science or religion, toward people who are seeking to know the truth or toward people who like hierarchies." Most yoga therapists get their information from a guru, he remarked. "That's what they believe and trust."

But science now has the means to determine what really works in yoga therapy and why, Fishman argued. Its methods can reduce false diagnoses and risky treatments. Its respect for the facts, he added, can help turn the fledgling discipline into a real profession.

VI

DIVINE SEX

In 1970, when I attempted my first Headstand, the topic of sex was typically relegated to the back room. My yoga books and those of my friends made few if any references to sexual aspects of the discipline. *The Complete Illustrated Book of Yoga* never mentioned Tantra, sexual arousal, or finding a willing "female partner," as *Hatha Yoga Pradipika* put it so charmingly. My first teacher made some passing remarks about sex. But I could never figure out exactly what he was talking about and left it at that for what turned out to be decades.

I began this book on the same note. Sex seemed sort of irrelevant. Oh, I figured it was out there somewhere and might produce a good chapter. But for the longest time I had no idea what would materialize and kept getting annoyed every time I dug into the scientific literature.

The studies were few in number and appeared to be contradictory and downbeat. One said yoga reduced the circulating levels of an important class of sex hormones. Intuitively, that seemed wrong. From personal experience, it seemed obvious that yoga stirred a number of hormones, some most likely sexual in nature.

But the scientific evidence seemed to point elsewhere. I did the easy thing and put the topic aside.

What eventually turned me around was the testimony of advanced yogis. I was amazed to find a new generation speaking with great candor about their autoerotic highs and how yoga by nature sought to recast the body for the purpose of sexual pleasure. In interviews, some spoke frankly of their bliss and endeavors to make the rapture permanent. An attractive yogini jokingly called her blinding ecstasies the best sex she never had. Some addressed the issue with what seemed like great reverence, saying yoga could turn the sexual experience into the holiest of sacraments.

I proceeded to discover that modern yoga throbs with open sexuality ranging from the blatantly erotic and the bizarrely kinky to the deeply spiritual. The veil hung ever so carefully by the early gurus and the Hindu nationalists has fallen away.

Routinely, yoga now promises to transport any serious practitioner into realms of sexual bliss that go far beyond the hot, moaning, knee-knocking variety of the bedroom. The trend is highly commercial in nature and has produced many thousands of books, websites, how-to articles, and video discs. *Better Sex Through Yoga*—a set of three DVDs (beginner, intermediate, and advanced)—promises to reward the student with "intense, long-lasting, full-body orgasms." How long? The woman teacher gives no particulars. But scantily clad and smiling coyly, she promises to take "you harder, deeper, and further than you've been in your workout—and your sex life."

After exploring this world for a while, the big picture suddenly came into view. I saw how limited science had obscured key evidence, why yoga reverberated with so many scandals, and how the discipline itself began as a sex cult. The pieces of the puzzle, as they say, fell into place.

One revelation centered on sexual misconduct among some of the world's most celebrated gurus. I learned of philanderers who acted with impunity and female victims who tended to rationalize the sex as some kind of spiritual test or ritual initiation. Most had a difficult time finding fault with men they saw as virtual gods.

Happily, my research also showed that the women began to resist and even take legal action. In 1991, protestors waving placards ("Stop the Abuse," "End the Cover Up") marched outside a Virginia hotel where Swami Satchidananda (1914–2002)—a superstar of yoga with long hair and a full beard who gave the invocation at Woodstock—was addressing a symposium. "How can you call yourself a spiritual instructor," a former devotee shouted from the audience, "when you have molested me and other women?"

Another case involved Swami Rama (1925–1996), the man who impressed scientists by seizing control of his palm temperature. In 1994, one of his victims filed a lawsuit charging that he had initiated the abuse at his Pennsylvania ashram when she was nineteen. He evaded deposition. Ultimately, he traveled to India, leaving behind his ashram in the Pocono foothills and its four hundred rolling acres. The case moved ahead despite his absence. In 1997, shortly after his death, a Pennsylvania jury awarded the young woman nearly $2 million in compensatory and punitive damages.

Even Kripalu came under fire. Former devotees at the Berkshires ashram won more than $2.5 million after its longtime guru—a man who gave impassioned talks on the spiritual value of chastity—confessed to multiple affairs.

I came to see these episodes as windows into the unruly forces at work in some of yoga's most developed bodies. The fallen seemed to confirm Iyengar's point about the crossroads of destiny. For science, the cases suggested that vigorous practice could stir the hormones and passions to such an extent that even pious men of high ambition could lose their way. The misadventures also offered a bittersweet tribute to yoga revitalization. It turned out that a surprising number of the philandering gurus were in their sixties and seventies.

My take on the subject kept getting reinforced as new episodes broke into public view—at times with a colorful new spin. Bikram Choudhury, the hot entrepreneur, a man known for libidinal energy and a love of hyperbole, was asked about rumors of having sex with students. The sixty-four-year-old guru offered no denials but claimed he was blackmailed. "Only when they give me no choice!" he exclaimed. "If they say to me, 'Boss, you must fuck me or I will kill myself,' then I do it! Think if I don't! The karma!"

With new resolve, I dug deep and uncovered a small trove of illuminating reports and investigations. They showed that yoga can in fact result in surges of sex hormones and brain waves, among other signs of sexual arousal. The newest studies add the weight of clinical evidence. Medical scans indicate that advanced yogis can shut their eyes and light up their brains in states of ecstasy indistinguishable from those of sexual climax. Meanwhile, new practitioners report that yoga improves their sex lives. The men and women say the benefits include better arousal, satisfaction, and emotional closeness with partners.

Little of this information is known publicly, despite yoga's reembrace of Tantra and the erotic. Most is lost in the labyrinth of modern science.

I have come to see the lack of understanding as not only a disciplinary weakness but something of a missed opportunity. Yoga practitioners may know from personal experience that the discipline can act as a potent aphrodisiac and revitalize their sex lives. But the professions of medicine, health care, and psychological counseling know little or nothing of such benefits despite their tireless promotion of costly treatments for low libido, arousal disorder, and sexual frustration. The same holds true of popular health guides.

As a result, sex authorities seldom if ever mention a holistic therapy that is quite natural and—as Fishman put it—free.

The ignorance goes right to the top. When Abraham Morgentaler wrote his 2008 book, *Testosterone for Life: Recharge Your Vitality, Sex Drive, Muscle Mass & Overall Health!*, the Harvard professor talked mainly of gels, creams, patches, injections, and pellets—all of which require prescriptions. His book made no mention of yoga, like most guides to hormone therapy.

The global pharmaceutical complex thrives on sex treatments, with sales booming in recent years. The marketing push is known derisively as Orgasm, Inc., and critics question whether it puts corporate profits above personal health.

It turns out that science over the decades has slowly uncovered an alternative that draws on the body's own hidden resources. It has no advertisements, no sales force, no hustle, no giveaways for doctors, and no questions about pressure to take unnecessary and possibly unsafe drugs. If nothing else, it seems worth investigating.

Katil Udupa was an ambitious physician at the Benares Hindu University, his professional life a blur of activity on the school's sprawling campus outside the holy city on the Ganges. He was, in many respects, a successor to Paul—a man of Western medicine who became deeply interested in the healing arts of India. He also exhibited some of Gune's passion for institution building. In 1971, Udupa founded the school's Institute of Medical Sciences.

He then proceeded to fall apart. After years of administration and its predictable crises, his outgoing nature started to crumble and Udupa ended up with a variety of nervous ills—chest pains, irritability, diffuse apprehension, emotional instability, and a sense of constant fatigue. The formal diagnosis was cardiac neurosis. We would call it burnout. Whatever the name, he was a nervous wreck.

Udupa took up yoga and found quick relief that slowly developed into a deep sense of personal renewal. Intrigued, he began to study the medical literature about yoga and to investigate its potential for treating patients—especially those with chronic diseases that appeared to be linked to the kind of stresses and illnesses that he himself had experienced. His studies showed that yoga could dramatically improve a patient's hormone profile, lowering, for instance, the high levels of adrenaline and other fight-or-flight hormones released in response to stress.

The body always puts survival ahead of pleasure. A corollary of that principle is that stress can smother the flames of desire, and relaxation can create a situation where smoldering embers get fanned into a blaze.

Udupa wondered if that kind of relationship held true on the biochemical level as well and, specifically, whether the reductions he was seeing in stress hormones meant that the body's sex hormones were tending to increase. It was a smart question.

He and his colleagues studied a dozen young men. Their average age was twenty-three, about half of them single, and half married. The volunteers underwent yoga training for six months. The lessons started out easy and, month by month, grew harder. The first month included the Cobra, the Spinal Twist, the Wheel, and the Full Lotus.

Spinal Twist, *Ardha Matsyendrasana*

New poses added over the months included the Plow, the Locust, the Bow, the Shoulder Stand, and the Headstand. The pranayamas included Bhastrika and Ujjayi, or Victorious Breath. Overall, by modern standards, the training was fairly rigorous.

The scientists took urine samples from the young men at the start of the program and its conclusion. They found that the urinary excretion of testosterone rose significantly, its levels in some of the married men more than doubling. On average, the levels rose 57 percent. The results, the scientists wrote in 1974, suggested that yoga could prompt a "revitalization of the endocrine glands." As for the mechanism, they speculated that yoga had improved the microcirculation of the blood through the men's organs. In males, testosterone is made primarily in the testes but also to a lesser extent in the adrenal glands. It seems that poses such as the Bow, which exerts pressure on the genital region, might well serve as a stimulus to improved circulation.

Bow, *Dhanurasana*

In 1978, Udupa published a summary of the hormone findings in his book *Stress and Its Management by Yoga*. He noted the clinical evidence of the testosterone rise and attributed it to "considerable improvement in the endocrine function of the testes."

His hunch had proved correct. But Udupa made little of it. His finding was a particle of basic science in a blizzard of global research.

• • •

Living peacefully on the Ganges a couple of hundred miles downstream from Benares and Udupa were advanced yogis who displayed a strong interest in science, their guru having named their ashram the Bihar School, after its location in Bihar state. It turns out they were interested in learning as well as teaching. A swami writing in *Yoga Magazine*, published by the school, called attention to Udupa's testosterone finding. His brief reference was buried in an overview of Udupa's yoga research. Still, the author, steeped in British English, noted how the hormone discovery suggested that yoga postures could improve "vitality and sexual vigour."

His appraisal was clear-eyed but rare. For the most part, science as well as popular and yogic literature ignored the finding. The Bihar yogis noted the testosterone rise in 1979, shortly after the publication of Udupa's book. It seems plausible that the finding caught their attention not only because of their proximity to the research but because their own experiences had convinced them of its physiological truth.

If science ignored the finding, investigators nonetheless threw themselves into acquiring a better understanding of testosterone. In Udupa's day, the potent hormone was seen mainly as the force behind the male sex drive. Scientists knew that its levels fell with age, and that rises could lead to revitalization.

But over the years, modern biology found many other ways in which the little molecule can influence behavior and sexuality—doing so in both males and females. Not the least significant, studies showed that it acts to improve mood and a person's sense of well-being. It seems likely that the hormone forms a significant part of yoga's cocktail of feel-good chemicals.

Importantly, testosterone was shown to bolster attention, memory, and the ability to visualize spatial tasks and relationships. It sharpened the mind.

Surprisingly, testosterone also turned out to play an important role in female arousal. While adult males tend to produce ten times more testosterone than females, scientists found that women are quite sensitive to low concentrations of the hormone. They make it in their ovaries and adrenals, and its production peaks around the time of ovulation—a phase of the reproductive cycle associated with increased sexual activity. A number of studies have linked testosterone rises in women to enhanced desire, erotic activity, intimate daring, and sexual gratification. The pharmaceutical industry is closely studying the hormone in hopes of finding a blockbuster drug like Viagra that it can sell to women.

Udupa's research got little attention not only because it was done in faraway India. Science for many centuries has been international in character, and back in the 1970s obscure articles could get quick attention if they revealed something bold and new. A factor that added to the disregard was the emergence of other studies that seemed to contradict Udupa's testosterone findings. Thus, scientists aware of the work increasingly saw his conclusions as hollow. In short, the topic became muddled—a common problem in backwater fields of science that fail to get the kind of intense scrutiny that can rapidly clear up complicated topics.

It was the low-testosterone findings that initially led me to conclude falsely that yoga had little to do with human sexuality.

The inconspicuous challenge to Udupa's findings grew out of an ambitious body of yoga research that sought to show how yoga could result in major benefits for cardiovascular health. It was the kind of thing that Dean Ornish had pioneered in the 1970s and 1980s, and that, over time, had received wide notice and emulation. In the research, scientists typically had subjects adopt not only Hatha yoga but other lifestyle changes as well, such as becoming strict vegetarians. In 1997, for instance, scientists at the Hannover Medical University in Germany reported on a study that examined more than one hundred adults who took part in a comprehensive program of yoga and meditation for three months. The setting was a yoga school that had its own fields, garden, and kitchen, and where the subjects adhered to its regular vegetarian diet. The results showed that participants lost weight and reduced their blood pressure and heart rates, significantly lowering their risk factors for heart disease. That was the study's main question, and scientists hailed the results as showing yoga's benefits for cardiovascular health.

But their report also noted that testosterone fell significantly. It was an aside—a minor finding in relation to the main question. But the idea nonetheless got lodged in the scientific literature.

Now, as it turns out, vegetarianism alone reduces the body's levels of testosterone, and that kind of reduction has been understood for a long time. The vegetarian factor meant that yoga most likely had nothing to do with the reported testosterone drops. Even so, the relationship had become cloudy. In addition, the muddle probably grew because of the structural bias in science that favors new findings over old. For whatever reason or reasons, fallacies gained currency.

"You won't boost testosterone doing yoga," Al Sears, a popular author and Florida doctor specializing in men's health, declared in a leaflet. "Try wrestling, boxing or karate instead."

The confusion meant that testosterone fell off the map for writers of yoga guides and how-to books. Its disappearance was understandable. At best, news of the hormone's reduction in the body was puzzling given the personal experience of revitalization and, at worst, seemed like something of an embarrassment. How could yoga do that given testosterone's importance for improving mood, attention, and sense of well-being—not to mention sex?

One way that popular yoga handled the ambiguous situation was by hailing the discipline's sex benefits while omitting any mention of testosterone. The 2003 book *Real Men Do Yoga* reported that the discipline "recharges your sex life" with "Viagra-like effects." But it made no mention of the potent little hormone.

Science kept inching forward, despite the jumble. In Russia, three decades after Udupa, investigators reported new evidence that echoed his findings. The team leader was Rinad Minvaleev, a physiologist who practiced yoga and led expeditions to the Himalayas. Among his interests was the Tibetan yoga of Tummo. It generates inner heat that is said to protect its practitioners from extreme cold. A photograph of Minvaleev in the Himalayas shows him sitting atop a glacier wearing nothing more than a swimsuit.

His team at Saint Petersburg State University and the city's Medical Academy of Postgraduate Studies undertook a yoga study with a very narrow focus. The subjects included seven males and one female, their ages ranging from twenty-two to fifty. The volunteers were taught how to do the Cobra and to hold it for two to three minutes. The team limited itself to studying the physiological repercussions of just that one posture.

The Cobra, or Bhujangasana, from the Sanskrit for "snake," is one of Hatha's oldest poses. It dates from the pure Tantra days before the era of sanitization and takes center stage in such works as the *Gheranda*

Samhita—a holy book of Hatha that scholars date to the transition between the seventeenth and eighteenth centuries. Parts of *Gheranda Samhita*, no less than *Hatha Yoga Pradipika*, read like a sex manual, full of references to the perineum, scrotum, penis, and so forth, as well as acclaim for the goal of stoking "the bodily fire." Bhujangasana is praised as an igniter. As the yogi performs it, the book says, "the physical fire increases steadily." The book describes the concluding step of the yogic journey as "pleasures, enjoyments, and ultimate bliss."

Easy to do, the Cobra is basic to beginning yoga and was one of more than a dozen poses that Udupa's subjects had performed. His volunteers began the Cobra in the first month of their practice, and thus did it longer than many of the other postures. The student, lying facedown, legs together, simply brings the hands forward and pushes down on the palms, raising the chest and head. Done correctly, the pose exerts much pressure on the genitals. As Iyengar puts it delicately in *Light on Yoga*, the pupil should lift the trunk "until the pubis is in contact with the floor and stay in this position." He adds that the student, once up, should "contract the anus and the buttocks," a move that increases the downward pressure.

In designing their experiments, the Indian and Russian teams took very different approaches and had very different goals in mind. The Indians looked at cumulative effects of yoga over six months, while the Russians looked at the repercussions of just one session. The Russian team simply drew blood before and after the volunteers did the Cobra, taking the samples no more than five minutes apart. It was a snapshot versus a movie. And because of the shorter period of training, the Russian results seemed preordained to show a more modest testosterone rise.

In their report, published in 2004, the Russians first told of changes they observed in levels of cortisol—a hormone that, as part of the body's reaction to stress and sympathetic stimulation, raises the blood sugar and blood pressure in preparation for an individual to flee or fight. On average, cortisol fell 11 percent.

As for testosterone, the team reported an average rise of 16 percent. Individual males showed increases that varied anywhere from 2 to 33 percent.

But the gold star for the biggest increase went to the study's lone female. Her testosterone readings shot above those of the males and kept rising to reach 55 percent—rivaling the increases that Udupa's male subjects had experienced after doing yoga for six months.

The Russian scientists, in their report, put their celebrity in the spotlight. The photo showed the attractive young woman clad in a bikini, rising into the Cobra, a picture of vitality and vigor. She almost seemed to glow.

Yoga as an exercise seems fairly distinctive in its ability to raise testosterone levels. Over decades of study, scientists have consistently found that endurance sports have just the opposite effect. Runners, for instance, show lower testosterone levels than nonrunners. The declines may result from the continuous stress of pounding the pavement.

Scientists looking into the relationship between yoga and sexual revitalization cast their gaze far beyond hormones and the body's endocrine system. In time, they zeroed in on a more central organ—the brain. Again, the research took place outside the United States, this time in Czechoslovakia. One of the main investigators held both medical and doctoral degrees.

Ctibor Dostálek fell for yoga in his early forties when he was already a skilled Czech neurophysiologist and longtime academic working in Prague. The year was 1968 and his deep personal interest altered the trajectory of his career. It sent him to India and Gune's ashram. His interest began with the sanitized version of the discipline but soon encompassed old Hatha. In all, he went to India eleven times. His research examined not Headstands and Sun Salutations but various kinds of stimulations out of the pages of *Gheranda Samhita* and *Hatha Yoga Pradipika*.

Dostálek's main tool was the electroencephalograph, or EEG. Compared to Gune's X-ray machine, it was all nuance and subtlety, giving a glimpse of firing neurons in action. He would cover the scalp of an advanced yogi with a dozen or more electrodes, switch on the machine, and peer into a hidden world. The EEG monitors faint currents of bioelectricity running across the brain and amplifies them roughly a million times, producing a graphic record of wavy lines. By 1973, Dostálek had become so proficient at electroencephalography that he was named director of the Institute of Physiological Regulations of the Czechoslovak Academy of Sciences. He worked in the heart of Prague.

Soon Dostálek turned his attention to his personal interest, yoga. One experiment focused on a single exercise—a kind of minimalism that reduced the influence of potentially confusing variables, much as the Russians had done in their Cobra study. The pose was Agni Sara, Sanskrit for "fanning the flame." Named after the Hindu god of fire, it had nothing to do with pop yoga or sleek gyms but instead arose from the misty world of Tantra. Modern gurus held the practice in such esteem that some recommended doing it daily even if time allowed no other exercise.

Doing Agni Sara properly can be difficult. The yogi leans over from a standing position, knees slightly bent, hands on thighs. After a deep exhalation, he or she holds the breath out and repeatedly pulls the stomach in and out. The objective is to tug backward toward the spine, then relax the stomach and let the abdomen fall forward. The cycle is repeated ten or fifteen times before the yogi inhales. The *Gheranda Samhita* tells students to do Agni Sara one hundred times. Swami Rama, the modern yogi known for feats and philandering, suggested doing it one hundred and fifty times daily.

Over the decades, science has learned a lot about the target of such undulations. The waves going through the lower abdomen massage the internal organs and nerves of the reproductive system. The area is often characterized as an erogenous zone. Masters and Johnson reported that, during sex, contractions of the abdominal muscles build into spasms that amplify the actions of pelvic thrusting.

Undulations higher in the abdomen massage the region devoted to digestion and its specialized nervous system. The complexity of the area is so great that scientists liken its network of nerves and neurotransmitters to a second brain. The system envelops the viscera in whorls of nerves and sensory receptors in order to exert control over the complex processes of digestion and elimination. To that end, it makes dozens of different hormones and neurotransmitters. Individuals can sometimes feel the subtle workings of the second brain as gut instincts, intestinal unease before a talk, or butterflies in the stomach. It can feel emotions and remember experiences. Stress and repressed feelings can upset its functioning, darken moods, and harm overall health.

For his experiment, Dostálek chose an elite subject, a disciple of Gune's who had done yoga for more than three decades. The thumpings of Agni Sara sent waves of stimulation rippling through his abdominal cavity, and the electrodes of the EEG revealed bursts of brain excitation. The spikes grew in size as the pounding intensified.

In a 1983 report, Dostálek called the peaks "very significant." The web of electrodes on the yogi's head let the electrophysiologist pinpoint the origin of the bursts. They arose from the central parietal lobes—a region of the brain responsible for processing body sensations, including touch and pressure. It is the parietal lobes that hold a miniature sensory map of the body whose anthropomorphic expression is known as the homunculus—a tiny human figure distorted to show the relative importance of sensory inputs to the brain. The homunculus has relatively big lips, hands, and genitals.

Dostálek widened his investigations to include more experienced yogis and more Tantric poses as well as fast breathing, including Bhastrika, or Breath of Fire, as well as Kapalabhati, or Shining Skull Breath. The blitz set off large bursts. Dostálek found that the spikes now built with greater force, cresting in frenzied excitation. In a 1985 report, he called the peaks "paroxysmal," in other words, like a seizure or convulsion. He noted that other scientists had previously observed such readings among people in sexual climax.

Dostálek pulled back at one point during his investigations to reflect on the big picture. He did so in the pages of *Yoga Mimansa*. In effect, his article was a thank-you to Gune's institution.

A misapprehension had arisen in science's understanding of yoga, Dostálek argued. A number of investigators and studies had come to view the discipline as strictly calming and relaxing, seeing its methods as producing "states of lowered autonomic arousal and EEG deactivation." But that was only half the picture, he argued. The other half told of excitation. His own studies, Dostálek said, revealed the magnitude of the stimulation. He noted that yoga's physiological arousals were associated mainly with the kinds of practices and fast breathing described in *Hatha Yoga Pradipika* and *Gheranda Samhita*.

Still, in his theorizing, Dostálek hewed to the conventional line. He portrayed the highs as simply helping the body gain new physiological flexibility, letting it not only rise to new heights but to hit new lows. Yoga worked in the main, Dostálek argued, to promote relaxation, unwinding the sympathetic overstimulations of modern life. Ultimately, he saw yoga as a sedative.

• • •

A few scientists glimpsed a different world. Their examinations of advanced yogis suggested that deep relaxation, rather than an end in itself, could represent a calm stage on the road to a remarkable kind of continuous arousal. Their subjects displayed clear signs of autonomic stimulation while lost in blissful trances. The studies were relatively few in number. But they were ample enough to suggest that, at least in some comparatively rare cases, the kind of fleeting arousals that Dostálek had documented could endure.

James C. Corby, a psychiatrist at the Stanford University School of Medicine, did the most thorough study of this hidden world. His team looked at twenty members of a Tantric sect known as Ananda Marga, or "path of bliss." The group, founded in India, treads a steep path. In addition to doing asanas, pranayama, and many austerities, the initiates meditate for long periods. The scientists recruited equal numbers of trainees and experts. On average, the expert Tantrics, all from the San Francisco area, meditated for more than three and a half hours a day and had done so for years. In their study, the Stanford team noted that practitioners often reported feeling rushes and bursts of energy during their meditations. The scientists also recruited ten inexperienced individuals to act as a control group.

Each subject sat alone in a dimly lit room during the monitoring sessions, which lasted an hour. The scientists had all the participants— whether controls, trainees, or experts—perform the same routine. The participant would spend twenty minutes relaxing, twenty minutes paying attention to their breathing, and twenty minutes meditating. The controls used two-syllable mantras they made up, while members of Ananda Marga used their personal mantras.

Corby and his team studied not only the brain waves of the participants but their heartbeats, breathing rates, and skin conductance. The latter was an important sign of emotional arousal and, in some cases, sexual excitement. Scientists have long known that sweat causes the electrical conductivity of the skin to rise, and have long viewed it as an indication of sympathetic arousal. In the early days of conductivity studies, scientists monitored the skin response as a way to probe the unconscious. So, too, scientists developed lie detectors as a way to measure skin conductivity for clues as to whether a person was relaxed and telling the truth or clammy and deceitful.

The Stanford team found that the Tantric experts and trainees displayed solid evidence of autonomic arousal. The signs included fast heartbeats and significant rises in skin conductance. The control group, on the other hand, showed signs of overall relaxation.

In one case, a woman Tantric sent the measurements flying off the charts when she experienced what she later described as "near samadhi"—the ecstatic state of enlightenment. While meditating, her skin conductance soared and she began to breathe fast and her heart rate shot up to more than one hundred and twenty beats per minute—equal to that of frenzied lovers. Abruptly, she stopped breathing altogether and her heartbeat slowed as well. Finally, after more than a minute and a half in which her chest remained virtually motionless, she began to breathe normally.

"We were extremely fortunate," Corby's team wrote, to observe the woman's experience.

Corby and his colleagues said nothing about sexual arousal. They used phrases such as "physiological activation" and "autonomic arousal." But their paper published in the *Archives of General Psychiatry*, part of the cautious world of the American Medical Association, based in Chicago—strongly implied the sexual basis for their findings given that the study's subjects were Tantrics.

To the best of my knowledge, this paper represents the closest that the scientific community ever came to identifying what I have come to think of as the yoga paradox—the sharp reversal in advanced yogis from physiological cooling to arousal, from states of hypometabolism to hypermetabolism. The paradox has nothing to do with the kind of false metabolic rise that Payne advertised and everything to do with one of yoga's biggest secrets.

Unfortunately, Corby's paper sank like a stone. The paper cited some of Benson's research and appeared while *The Relaxation Response* was still popular and well on its way to selling millions of copies. For many years, the relaxation paradigm continued to dominate the scientific concept of how yoga worked. The alternative perspective that stressed rare states of continuous arousal—for a variety of reasons—stayed in the shadows.

• • •

As Udupa looked into hormones, Dostálek into brain waves, and Corby into skin conductivity, other scientists were examining a rather curious but poorly understood parallel between yoga and sex—heavy breathing. Dostálek got a glimpse of the similarity when his Bhastrika subjects felt "elation and even exhilaration." So did Corby when his Tantric meditator soared toward samadhi. But brain waves and skin conductivity were just two of many ways to explore the repercussions of rapid breathing. Indeed, its basic study required no specialized equipment at all. The most fundamental method was just to sit quietly and watch.

In *Human Sexual Response*, Masters and Johnson describe fast breathing as an integral part of male and female behavior leading up to sexual climax. The scientists reported rates of more than forty breaths per minute at the height of strong orgasms. Compared to normal rates of relaxed breathing, that is roughly three times as fast.

The pace of heavy breathing during sex may seem rapid but it is nothing compared to aggressive Bhastrika. Yoga teachers tell beginners to start at one breath per second and work their way up to two breaths per second—or one hundred and twenty breaths per minute. Advanced students are encouraged to take up to four breaths per second. If done without pause, that equals two hundred and forty breaths per minute—a rate five or six times faster than lovers.

As we saw in chapters 3 and 4, heavy breathing can pose significant risks of injury and even death. But if done in moderation, it can be quite benign. Mild hyperventilation does no permanent damage to the brain or the nervous system but simply contributes to the sense of euphoria that makes both sex and yoga so enjoyable.

Over the decades, scientists worked hard to explore sexual hyperventilation, usually in the interest of understanding Western sex rather than Eastern asceticism. Still, the overlap was great enough so that some investigators argued that the insights applied to both.

More recently, a body of emerging research has revealed that fast breathing can not only lower the flow of oxygen to the brain, as we saw in chapter 3, but sharply reduce activity specifically in its outer layers. The findings grow mainly out of the technology of brain scanning. It lets scientists peer deep—going far below the superficial regions that researchers had explored with the electroencephalograph—to compare the levels of inner and outer activity.

The scanning unveiled a primal experience that amplified the body's surges of pleasurable hormones and brain waves.

In biology, the outer brain is known as the cortex. The term confuses many nonspecialists because it has nothing to do with a core. It's about edges and coverings. The term derives from the Latin word for "bark," as in the bark of a tree. The cortex plays important roles in memory, attention, calculation, awareness, thought, empathy, abstract reasoning, language, and sensations, as we saw with the parietal lobe. Right now it is interpreting these words. It also appears to be the seat of consciousness. The area known as the prefrontal cortex ("pre" because it is the most forward part of the brain, way out front, right behind the forehead) is well developed only in primates, especially humans. It controls such higher functions as planning, decision making, and setting priorities.

Deeper down is the older, more primitive brain. Here lie raw appetite and unfettered passion. The deep structures of the primal brain include the neuroendocrine system, made up of the pineal body, the pituitary gland, and the hypothalamus, with its vigilance area and its control of the autonomic nervous system. Another cluster makes up the limbic system. It wraps around the brain stem and supports such functions as emotion, motivation, homeostasis, and short-term memory.

The limbic system also controls sex. The amygdala, a limbic body made of two lobes about the size of almonds (its name comes from the Greek word for "almond"), plays major roles in emotion, including aggression and pleasure. As for sexuality, it has the brain's highest density of receptors for sex hormones, including testosterone. Scientists have shown that the stimulation of the amygdala results in a wide variety of sexual activities, including erection, ejaculation, ovulation, and the rhythmic movements of copulation. The stimulus can be purely hormonal. Dutch scientists recently studied middle-aged women whose amygdalas had undergone decline and found that small doses of testosterone could restore the organs to youthful vigor.

As the scans let scientists peer deep, they began to see that fast breathing had different repercussions on different parts of the brain. The limbic system experienced no activity drops like those of the cortex. It moved to its own beat. Sex and hyperventilation could deprive the cerebral cortex of blood and oxygen, diminishing the higher functions of the brain, even as its inner regions kept going strong. It was like a late night in the suburbs. The residents turned out the lights upstairs while continuing to party in the basement.

In Germany, Torsten Passie, a psychiatrist at the medical school of the University of Hannover, drew on the limbic findings to propose a theory of sex hyperventilation. The decrease in cortical management, he wrote, resulted in a "more primitive mode of brain functioning" marked by heightened emotions, declines in self-control, and a deepening sexual trance.

All of which led to an intriguing question. Could fast breathing from yoga or anything else produce a sexual high in and of itself? The question went to knotty issues of causation. Was fast breathing solely a result of sexual stimulation, or could it also work as an initiator?

In Vancouver, Lori A. Brotto and other sex researchers at the University of British Columbia began looking for answers. The scientists recruited twenty-five women—all heterosexual and sexually experienced—and measured their responses to an erotic film. The reactions were noted twice, once after hyperventilation and on a different day without the benefit of fast breathing. The women took thirty deep breaths per minute for two minutes. By the standards of Bhastrika and other kinds of rapid yogic breathing, the routine was fairly mild. Even so, the researchers judged that the breathing produced a state of sympathetic dominance that lasted at least seven minutes.

Their report, published in 2002, showed that the women watching the film experienced a significant rise in the amplitude of their vaginal pulses, suggesting that viewing the amorous film did in fact produce genital arousal. As a group, the amplitude doubled.

That led to an investigation of whether the breathing technique could have practical applications. Brotto and her colleagues recruited sixty women with sexual arousal disorder, or SAD, and a control group of forty-two women with healthy sex lives. Again, the volunteers were all heterosexual and sexually experienced. The women saw two erotic films in a row. One group hyperventilated before the first film, and then had a resting period before viewing the second. The procedure with the second group was reversed, so fast breathing took place before the second film.

The results suggested that even short periods of fast breathing could improve arousal. As before, the healthy women responded more vigorously to the erotic film if they hyperventilated in advance. But so did the SAD women. Their histories included the absence of or diminished ability to respond to physical stimulation of the genitals as well as to visual and auditory cues that normally result in aroused feelings. In many respects, these women were difficult cases. Yet their levels of excitement were almost as high as those of their peers.

In short, the evidence suggested that hyperventilation could promote arousal not only in healthy women but among those with diminished libidos.

Fast breathing fanned the flames.

My first visit to a Bikram studio drove home the primal nature of yoga breathing. I had done yoga for decades. But now, in my new frame of mind, I realized that most classes were structured—by accident or design—to echo one of the most basic of all human experiences.

It was a full house that night, the mirrored room packed with men and women, most in great shape and apparently enthusiasts who practiced a lot. A man up front kept a big jug of water next to him. He clearly understood better than I how hot yoga could produce torrents of sweat and the urgent need for liquid refreshment (a situation not unknown in adult relations). No one seemed overweight. The group in general looked fit and attractive.

Bikram classes follow a routine of twenty-six poses that start and end with pranayama. Our first breathing exercise was slow and calming—good for warming up and helping beginners feel at home. By definition, it was hypoventilation that gently pressed the parasympathetic brake, relaxing body and mind. I felt warm and calm and aware, ready for anything.

The postures began easy and grew more challenging, as was usual for a yoga class. The bending and stretching got deeper and more pronounced, the tensions rising slowly. Sexologists describe growing muscular strain as an integral part of the human sexual response. The contractions start gently in the arousal phase and develop into tensions and flexions that are quite pronounced in the plateau phase—the time of extreme activity just before climax. So, too, we performed the hardest poses toward the session's end, pushing ourselves, stretching and straining, bathed in sweat.

The final breathing exercise was very fast. It was Kapalabhati, the relatively mild form of Bhastrika. To me, it was good old hyperventilation and sympathetic arousal, with the usual buzz and, after we finished, a sense of calm elation. Sexologists call it the resolution phase.

We lay on our backs in Savasana as the instructor dimmed the lights.

Yoga classes—with their bending, sweating, heavy breathing, and various states of undress—have acquired a certain reputation. *Sex and the City* cast the issue in graphic terms. Samantha in one episode gets so hot and bothered that she puts the make on a nearby guy. Rebuffed, she tries another and wins his enthusiastic nod, after which they hurry out of the room.

The show invented a term to describe the union of yoga and orgasm—*yogasm*. An ad campaign quizzed readers on the definition. One: a yo-yo trick. Two: sex with Yogi Berra. Three: what Samantha has with a guy from her yoga class.

The word entered the zeitgeist. In 2009, *The New Yorker* ran a cartoon showing a woman reading in bed next to her husband. "Not tonight, hon," the woman said. "I had a yogasm in class."

Vikas Dhikav was interested in whether yoga could not only arouse individuals but improve the sex lives of couples. In New Delhi, the young doctor assembled a medical team and more than one hundred male and female subjects. Dhikav and his colleagues published two papers in 2010. The results went far beyond the hints contained in decades of physiological research—not to mention the cartoons and videos, scandals and lawsuits, tales and testimonials. The clinical evidence argued that yoga did in fact have a talent for promoting intimacy.

The medical team asked the men and women to report on their sex lives before and after practicing yoga for three months. The poses of the routine differed slightly from the usual composition. The scientists chose postures for what they called their potential to improve "muscle tone, gonads, endocrines, digestion, joint movements, and mood." Although the team made no reference to the arousal studies discussed in this book, the pose selections turned out to include a number that those reports had identified as sexually stimulating. The poses included the Bow, the Wheel, the Plow, and the Locust (all from Udupa's study), the Cobra (from the Russian study), and Agni Sara (from Dostálek's study). Other poses included the Triangle and the Seated Forward Bend. The pranayamas included Kapalabhati, the fast breathing we did at the Bikram studio. In usual fashion, the subjects ended their sessions with the Corpse and relaxation.

The results sang. The novice yogis told of improvements in all categories of sexual experience under investigation—including desire, arousal, orgasm, and satisfaction. The men, age forty on average, reported enhanced abilities to maintain an erection during intercourse and increases in their degree of hardness. They also expressed greater confidence.

The women told of newfound excitement. Their ages ranged from twenty-two to fifty-five. As a group, they reported improvements across all measured categories, including several indicators of heightened pleasure as well as emotional closeness with lovers. The scientists also found that women at different life stages differed in what they considered the best results. Women over forty-five reported that the biggest gains centered on enhanced arousal. In contrast, the younger women reported that the largest improvements had to do with the quality of their orgasms.

The natural history of the human orgasm is a subject on which science has shed some light. Over the decades, teams of investigators have measured its length and discerned a well-defined experience that can vary considerably in duration and character. The usual range falls between a few seconds and twenty-two seconds. Masters and Johnson discovered that, in rare instances, certain women could experience orgasms that lasted a minute or more. They coined a fancy term for the situation, calling it *status orgasmus*. The *status* implied a continuous state rather than brief interlude. The scientists found that women experiencing such episodes appeared to move with extreme rapidity between successive orgasmic peaks, as indicated by repeated contractions of their vaginal walls. The measurements of one woman showed her undergoing more than two dozen rapid contractions.

Not surprisingly, the nervous system turned out to orchestrate the arousals. The most important shift featured the change from parasympathetic to sympathetic dominance. The parasympathetic—the rest and digest part—began the activity by promoting a state of relaxed engorgement and erection. In this phase, the reproductive organs of both males and females filled with blood. Then the sympathetic part of the autonomic nervous system would kick in, pumping adrenaline and throwing the body into a rising frenzy of tension, breathing, and pounding activity, as well as soaring heart rates and blood pressure. The sympathetic peak came at climax.

In exploring this world, science found a remarkable class of women who can *think* themselves into states of sexual ecstasy—a phenomenon known clinically as spontaneous orgasm and popularly as *thinking off.* At Rutgers University, scientists looked at ten women who claimed such abilities. Each was examined separately. In the laboratory, the scientists would have each woman lie down on a hospital bed full of decorative pillows, measure her excitement, and compare her response to readings generated when she stimulated her genitals manually.

The results were unambiguous. The scientists found that both conditions produced significant rises in blood pressure, heart rate, and pupil dilation (all due to sympathetic arousal) as well as tolerance for pain—what turns out to be a signature of orgasm. Some of the women, the scientists noted, "showed vigorous muscular movement" during their nongenital arousals while others "appeared to be lying still." The overall findings, the team wrote in a 1992 paper, called for "a reassessment of the nature of orgasm."

Significantly, yoga played a central role in developing some of these talents. One of the women was a yogini who was happy to demonstrate her abilities for the sake of science. She said she could focus on her spinal column and rapidly throw its energies into action. "Just tell me which chakra you'd like to measure," she told the scientist in charge. "I can orgasm up and down all the energy centers. I don't know how much time you've got, but I won't have any problem keeping things going all afternoon."

At first glance, the idea of experiencing sexual bliss over the course of hours, days, or a lifetime seems absurd. If regular orgasms involve the fleeting loss of contact with reality (what is sometimes known as *la petite mort*, "the little death"), then a rapturous experience that went on continuously would seem to leave its beneficiaries cut off from the world and permanently adrift. How would you eat, play soccer, or run a meeting? The idea of existing in both worlds simultaneously seems like a logical contradiction.

The objective may appear somewhat less dubious if you take into account the long intermingling of mysticism and sexuality. Across ages and cultures, the aims of the two have proved to be remarkably similar, if not identical. Both encourage states of single-mindedness. Eastern religions such as Taoism, Hinduism, and Buddhism all teach the mutuality of spirituality and sexuality. Christian ascetics also evoked the union. They often spoke of the soul, or "the bride," as seeking assimilation with the beloved.

Any visitor to Rome who has gazed on Bernini's *Ecstasy of Saint Teresa* sees a moving portrayal of this kind of spiritual euphoria. The saint's head is thrown back, her lips parted in what looks like erotic anticipation. You can almost hear her moan.

Not that living in two worlds is easy. In India, individuals caught up in ecstatic trances often have devotees who help with the basics of survival. Ramakrishna (1836–1886), one of modern Hinduism's great saints, kept assistants on hand to tell him when he had eaten enough. The saint could also forget to breathe. At night, devotees would stand watch and wake him if necessary.

In yoga, the path to continuous bliss is known as *kundalini*, although no one would fault a casual observer for thinking otherwise. The word tops my list of yoga's most confusing terms. First, kundalini refers to both a common variety of yoga as well as one of the discipline's most esoteric experiences (which the style targets). Advanced yogis tell me that perhaps 1 percent or fewer of all practitioners undergo kundalini arousal. But its audience is much larger. Public discussions of the phenomenon evoke all kinds of allure—knowledge, power, mystery, excitement, danger, ecstasy, and more—even while cloaking the blissful state in misapprehensions and euphemisms. My college dictionary does a fair job on the fundamentals while avoiding any hint of its underlying sexual nature: "In yogic tradition, spiritual energy that lies dormant at the base of the spine until it is activated and channeled upward to the brain to produce enlightenment."

The Sanskrit definition of kundalini is "coiled" or "she who is coiled," as in a coiled snake. That is the iconic representation. The serpent lies sleeping at the base of the spine and its uncoiling or awakening and movement up the spine is said to mark the beginning of enlightenment. The symbolism may seem odd. But the snake has a long history as a representation of rebirth because of its ability to shed its skin. In Hindu religious life, snakes enjoy high status and are often worshiped as gods and goddesses. So the traditional image of kundalini makes sense in terms of its cultural origins. The rising snake marks a new beginning. Of course, serpents have very different associations for readers of the Bible. It is no surprise that, in recent years, some evangelicals have assailed kundalini as the work of the devil.

The sinuous depiction is rooted, at least partly, in sensation. Ramakrishna said he sometimes felt the mystic current rising "like a snake" up his spine, the movement going "in a zigzag way."

The awakening of kundalini is also said to result in fiery sensations, its path through the body described as burning hot. In his treatise on yoga, Eliade, the historian of religion, cited ancient texts referring to kundalini as a "great fire" and a "blazing fire." It has, in short, been portrayed repeatedly as a kind of living flame. The etymology of the word reinforces that image. Its Sanskrit root, the verb *kund*, means "to heat or burn."

Tantric authorities describe the mystic fire as divine in origin and feminine in character, calling her a sleeping goddess that the accomplished yogi seeks to awaken. Her names included Shakti and Isvari, the goddess of supreme reality. The cosmic female element is said to surge up the spine to the top of the head and unite there with her male counterpart, Shiva, their communion producing a state of transcendent bliss.

Old accounts tend to be vague in describing the physical basis of kundalini. Modern depictions are no better. The definitions include mystic energy, enhanced flows of prana, the vital energy behind spiritual growth, and the mothering force that guides human development.

Yogani, an American Tantric who writes under a pseudonym and often makes references to modern science, rejects such portrayals as cover stories. His 2004 book sums up his perspective in a blunt chapter title: "Kundalini—A Code Word for Sex." He calls the mystic experience "a flowering of orgasm, an expansion of orgasm into endless full bloom in the whole body."

• • •

The main investigators of kundalini in the world of science turn out to have been not sexologists or biologists but psychologists and psychiatrists. The group is fairly small and typically works on the fringes of the therapeutic world. Moreover, it has achieved nothing like accord on whether the hotwiring of the human body is good or bad, healthy or pathological. Instead, the experts typically clash.

Remarkably, one of the first investigators—if not *the* first—was no less a figure than Carl Jung (1875–1961), the Swiss psychiatrist. He came upon a case of kundalini arousal early in his career and developed a deep interest. Around 1918, a woman of twenty-five came to his attention whose symptoms included a wave of physical turmoil that rose from her perineum, to her uterus, to her bladder, and eventually to the crown of her head.

He was baffled—and she delighted. "It's going splendidly!" the woman said of their analytic sessions. "It doesn't matter that you don't understand my dreams. I always have the craziest symptoms, but something is happening all the time." To Jung's astonishment, he realized belatedly that the woman found the physical and psychological chaos to be enjoyable.

Jung lectured repeatedly on kundalini over the years and in 1932 gave four talks in Zurich on its psychology. He endorsed its academic study but warned people away from its practice. One of his sternest admonitions came in 1938, two decades after taking in his kundalini patient.

Jung called the experience a "deliberately induced psychotic state, which in certain unstable individuals might easily lead to a real psychosis." The term is one of the darkest of psychiatry. It bespeaks serious breaks with reality marked by delusions, hallucinations, and other crippling failures of consciousness.

Kundalini, Jung concluded, "strikes at the very roots of human existence and can let loose a flood of sufferings of which no sane person ever dreamed."

The analytic tone changed dramatically in the 1970s as waves of Indian gurus swept the United States and many yogis and spiritual seekers began to undergo kundalini arousal. Lee Sannella (1916–2010) gave one of the earliest and most upbeat assessments. A graduate of the Yale medical school, the San Francisco psychiatrist led early seminars at the Esalen Institute, the icon of the human potential movement that explored drugs and sex, religion and philosophy.

For Sannella, the question was whether the mystic fire led to genius or madness, or some ambiguous mix of the two. His 1976 book *Kundalini: Psychosis or Transcendence?* told of thirteen people who had undergone arousal. They included an actress, a psychologist, a librarian, a professor, a writer, two artists, two housewives, a healer, a secretary, a psychiatrist, and a scientist. His portraits were anonymous.

Sannella said his survey indicated that kundalini represented no jump off the cliff but rather "a rebirth process as natural as physical birth. It seems pathological only because the symptoms are not understood in relation to the outcome: an enlightened human being."

Scholar that he was, Sannella did mention Jung, who by that time had become a counterculture hero because of his embrace of the mystic East. But Sannella downplayed the warnings. He devoted one sentence to Jung's conclusion that kundalini could lead to madness.

Sannella's case studies tended to follow the same script—initial difficulties followed by slow recoveries so that the awakenings ended on a happy note, with the individual feeling a deep sense of personal renewal. But the evidence suggests that he engaged in a considerable degree of interpretative spin. For instance, his portrayal of the Reverend John Scudder, an Illinois psychic healer, reads nothing like the minister's own account.

Scudder told of his body filling with heat, light, and energy. His blood seemed to boil. His organs felt like they were on fire. Waves of energy pounded his head. His heart beat so violently that alarmed friends could hear it thumping loudly in his chest, and their church later that day announced that he had suffered a heart attack. Sleep eluded him. Weeks of agony left him fearing for his life and his sanity, even as he judged himself able to read minds and see with clairvoyant vision. Then, quite suddenly, the horror ended and he felt thoroughly clean in a way he had never felt before.

Afterward, Scudder told anyone who would listen that the experience was to be avoided at all costs. "I was led to believe that the opening of the kundalini was a great and glorious occult experience," he recalled. "What I went through was absolute hell. If there is a hell, it could not be any worse than what I endured."

By the 1980s, aggressive gurus and practices had bestowed upon the San Francisco region many hundreds of kundalites, as students of the inner fire are known. Sannella alone came across nearly one thousand cases and helped found a counseling service known as the Kundalini Crisis Clinic. The Spiritual Emergency Network—later renamed the Spiritual Emergence Network for a more positive spin—did no counseling but ran a hotline. Between 1986 and 1987, it answered more than five hundred calls. An analysis showed the typical caller to be a woman, age forty, who had questions about kundalini.

Today, scores of websites around the globe offer advice, most hailing the fiery experience as a sure path to spiritual uplift. But some tell of terrors, of strange illnesses and life upheavals, of desperate visits to doctors who find it hard to imagine what is going on, much less what kinds of treatments to recommend. A few tell of heart attacks and even death.

Bob Boyd of Greensboro, North Carolina, founded a website known as *Kundalini Survival and Support*. There he told of his own arousal as a young man and the nightmare of being unable to extinguish the mystic fire. The blinding rushes, he wrote, "literally crippled me mentally in terms of what academic achievements and future accomplishments I may have had." People around the world, Boyd said, "rue the day they walked into the kundalini ring of fire."

Such warnings get little play while popular portrayals tend to gain wide audiences. Elizabeth Gilbert, author of the runaway bestseller *Eat, Pray, Love*, paints an alluring picture of her own experience at an Indian ashram. "I suddenly understood the workings of the universe completely," she gushes in her book. "I left my body, I left the room, I left the planet, and I stepped through time." Back on earth, she discovered that kundalini left her "randier than a sailor on a three-day shore leave."

A few entrepreneurs have seized on the raw eroticism as a way to turn a profit, moving from the austerities of yoga to the garishness of commercialism. Their products focus not on full-blown kundalini but on an assortment of lesser arousals that seem to have little to do with mysticism or healing. It's mostly about hedonism. Not surprisingly, California—home of distinction in the pursuit of drugs, sex, and other diversions—started the trend and became a hotspot.

A pioneer was More University. Founded in 1977, it flourished in the dry hills east of San Francisco, offering doctoral degrees in such subjects as sensuality. It had no library and no campus other than people's homes, yet a doctoral degree cost about fifty thousand dollars. What hundreds of students did learn was how to lengthen their orgasms. Graduates of the university reported one experiment in which a woman kept going for eleven hours. California, facing growing federal pressure to shut down diploma mills, eventually withdrew More's certification.

But the knowledge spread. A principal medium was how-to books, several by More alumni.

Patricia Taylor graduated from Barnard College in Manhattan and received a master's degree in business administration from the Wharton School of the University of Pennsylvania. She worked on Wall Street before transferring to San Francisco, where she studied Tantra. In 1988, her life changed when a More alum brought her into a state of ecstasy that lasted about twenty minutes. "I was breathing fire out of my hands and feet," she told me. "Then I went into the light."

After studying at More, she refocused her life on teaching how to achieve long orgasms, calling them "a portal to the divine." In 2002, she authored *Expanded Orgasm.* Her website, www.expandedlovemaking .com, offers books, advice, and courses, including intensives for partners.

Taylor told me she has been happily married for two decades. Her longest orgasm? Two or three hours, she replied. She added that it was hard to say exactly because it was easy to lose track of time.

Science is turning a new generation of imaging machines on these uncommon states in an effort to learn more about their characteristics and better understand the human sexual experience.

A pioneer is Barry Komisaruk, one of the first scientists to look into the neurophysiology of orgasm. The Rutgers professor worked with two female colleagues to publish the think-off study in 1992, and over the years has sought to map the neural aspects of sexuality, writing more than one hundred papers. His long interest resulted in an understated book, *The Science of Orgasm,* published in 2006 by Johns Hopkins University Press. By then, Komisaruk was not only doing research and teaching but was named associate dean of the graduate school.

Relatively late in his career, Komisaruk began using a new means of investigation that went far beyond the EEG in revealing how orgasms light up the brain. The technique, known as functional Magnetic Resonance Imaging, or functional MRI, showed changes in cerebral blood flow and thus neural activity. By the 1990s, functional MRI had come to dominate the world of brain mapping because of its easy operation, wide availability, and clear data. Its pictures showed the overall brain in grayish tones and areas of heightened activity lit up in oranges and yellows.

From his laboratory in New Jersey, Komisaruk began using the machine in the late 1990s to better understand the workings of neurophysiology and orgasm. By 2003, still fascinated by the think-off women of more than a decade earlier, he began a new round of experimentation meant to explore what functional MRI might reveal about their spontaneous orgasms as well as fundamental aspects of human sexuality.

Much good science gets done by eliminating the jumble of confusing variables that surround most aspects of nature. That is what Dostálek and the Russians did in examining the physiological repercussions of a single yoga pose. Komisaruk was attracted to the think-off women for the same reason. Spontaneous orgasms seemed to represent the human climactic experience shorn of the confusing variables of sensory input and muscular contraction. For brain imaging, that meant the sensory and motor cortex would stay grayish, as would most other regions of the brain normally involved in the human interaction with the external world. In theory, the functional MRI would show the purely limbic parts of the experience. Of course, women having orgasms without touching themselves might eventually shudder with pleasure, as the 1992 study had shown. But the commotion might start relatively late in the arousal. In theory, the new line of experimentation promised to produce what Komisaruk called a "cleaner picture" of orgasm and an opportunity to better understand its nature.

In 2003, upon examining the first images, Komisaruk was pleased to see confirmation of the study's conclusions from a decade earlier. The pleasure centers of the women's brains lit up more or less identically whether they reached their orgasmic highs by means of physical stimulation or simply thinking off. Different paths led to the same outcome.

The challenge was getting enough volunteers. A good study would require a fair number of subjects—all of them possessing a rare talent largely unknown to the world at large. The recruiting job required a light touch, good connections, and a bit of astute salesmanship. After all, what woman was eager to lie down on a hard table under the glare of fluorescent lights and have her head zapped by a giant donut-shaped magnet while attempting to let go?

It was a difficult proposition at best—difficult, that is, until Nan Wise came along. Nan Wise is an attractive sex therapist and yoga teacher whom Komisaruk got to know when she went back to school at Rutgers after raising two children. Studying yoga and learning how to pay close attention to the energy currents in her body had turned her into a skilled practitioner of thinking off, and she agreed to a functional MRI scan when Komisaruk asked her. "It's the least sexy thing in the world," she told me. "But I do it for science."

By early 2010, Komisaruk and Wise had succeeded in doing preliminary scans on half a dozen volunteers. Head movement turned out to be a significant issue. The orgasms that Wise herself experienced while in the machine had resulted in virtually no head motion and thus very clear images. But other think-off women often thrashed about. In one case, Wise recalled, "it looked like the scanner was going to jump around the room." As a solution, the scientists devised a head restraint that was bolted onto the machine. It worked. Now the heads of the think-off women held steady even if their bodies became agitated.

Wise decided to pursue the inquiry as part of her doctoral research. What she and Komisaruk envisioned was documenting the steps by which various neural circuits and networks lit up in orgasm. In essence, they wanted to make a brain-scan movie, hoping it would throw light on fundamental riddles. For instance, the research might help scientists learn how to distinguish the parts of the brain that mediate pain and pleasure. The brain in a state of orgasm, Wise told me, looks much the same as when it experiences pain. "We don't understand very much about what constitutes the difference."

For her dissertation, Wise needed at least a dozen think-off volunteers. But now, with the rise of Neotantra and alternative sexuality, recruitment in the New York City area proved to be easy. Wise knew her way around the sex-and-spirituality crowd and knew the right people to contact for volunteers. "I know somebody who knows somebody," she mused. "That's how it works." One group she drew on was One Taste. Its founder had taken up the methods of More University and set up businesses in San Francisco and lower Manhattan that promoted open sexual relationships as well as orgasmic meditation. Wise's think-off volunteers ranged from New Age mystics to radical feminists who preached the virtues of learning how to achieve sexual satisfaction without men.

The more Wise learned, the more she marveled at the diversity of euphoric states. "There are orgasms and there are orgasms," she said. "For me, thinking off feels like a diffuse orgasm. Now that I've been interviewing people who have this capability, some of them have unbelievably intense orgasms. I think some people can cue their nervous system in that direction pretty easily."

I asked about length.

"We've seen all sorts of different styles," she replied. "There seem to be some people who can create an orgasmic state and keep it going. I've never timed it. But there are people who can go on and on."

While science over the decades has made some progress in illuminating the relationship between sex and yoga, it has cast less light on an esoteric issue that is even more fundamental and important. For ages, the topic was seen as having to do almost exclusively with divine inspiration. Today, it is perceived as the heart of what it means to be human.

VII

MUSE

Paul Pond wanted to know how the universe began. His doctorate in particle physics from Northeastern University in Boston opened the door to a world of thinkers who sought to identify how the building blocks of nature coalesced in the first instants after the Big Bang, how things like mesons took shape and disappeared in bursts of other elementary particles. He published in *Physical Review*—the field's top journal—and did research in such places as Toronto and London, Paris and Vienna.

Then he began to undergo kundalini arousal. In 1974, he decided to give up physics research.

Pond and his friends lived in Canada, mostly in and around Toronto. But they became enamored of a Kashmiri mystic by the name of Gopi Krishna who lived half a world away. Late in the summer of 1977, Pond, along with more than two hundred and thirty other Canadians, boarded a jumbo jet and flew to India to visit the aging kundalite. A few helped him spread his message. In turn, the pandit visited Toronto in 1979 and again in 1983, a year before his death. Krishna shunned guru status. But the Canadians revered him as a visionary and felt an obligation to keep his agenda alive, most especially his passion for studying how kundalini could foster intuition and genius, insight and creativity.

Krishna taught that the mystic fire "must" turn a common person into "a virtuoso of a high order, with extraordinary power of expression, both in verse and prose, or extraordinary artistic talents." His teachings—laid out in *The Biological Basis of Religion and Genius*—made the human potential movement seem like a tea party.

A farm in southern Ontario became the headquarters from which Pond, his wife, and their friends spread the word. In 1986, they held the first of what would become decades of annual conferences under a big tent. They called their group the Institute for Consciousness Research. The small Canadian charity with the esoteric agenda became a magnet for hundreds of people. It sold kundalini books, built an extensive library, put out a newsletter, and sought to show that the mystic fire could result in artists and writers, saints and innovators. Over the years, it examined such figures as Brahms, Emerson, Gandhi, Victor Hugo, Thomas Jefferson, Walt Whitman, Rudolf Steiner, Saint Hildegard, and Saint John of the Cross. The published results were typically rich in endnotes.

Pond underwent his own transformation. He became more open to people. So did his writing. As a scientist, he had specialized in papers that were extremely dry. Now he found pleasure in poetry—something he had previously avoided and engaged in only when forced to do so in school. The muse compelled him to write.

Restless ego like a child
   eating candy, running wild.
I say ideas come from a higher source
   but secretly wish they're mine of course.

Life holds few mysteries greater than those concerning the wellsprings of creativity. Thinkers down through the ages have developed many theories about what keeps the springs flowing and what causes them to dry up. Freud proposed one of the most enduring when he suggested that the sublimation of sexual energy fosters the artistic temperament and the creative impulse. But he denied that he, or psychoanalysis, could provide much else by way of explanation. "Before the problem of the creative artist," Freud remarked in a study of Dostoyevsky, "analysis must, alas, lay down its arms."

Despite the durability of the question, a fair body of evidence—much of it anecdotal, some of it middling, parts of it robust—has emerged over the decades to suggest that yoga can play a role in stirring the wellsprings. And kundalini is only part of the story.

The evidence has accumulated even though the issue is scientifically challenging. Creativity, after all, is rooted in human subjectivity, and even the best investigators can have a hard time finding ways to explore the ephemeral nature of inspiration. By definition, the research is much more difficult to do properly than measuring hormones and muscle tension, brain waves and blood pressure.

A complicating factor is that the overall issue of yogic creativity tends to be poorly known. It has received little public attention compared to more popular aspects of the discipline. The low profile and lack of buzz mean that scientists face serious challenges in trying to obtain funding to pursue the unfamiliar lines of research.

Even so, the topic is potentially quite important. Artists and creative thinkers have reputations as rebels. But throughout history, they have starred not only in the annals of invention but in the social upheavals that frequently result in periods of civil progress. If yoga contributes to the advance of artistry, it seems like the discipline might act as a cultural force of some consequence.

This chapter explores that possibility and the extent to which science in its current state of development can illuminate the topic.

The potential links between yoga and creativity often lie hidden in plain sight. For instance, Carl Jung relied on the calming effects of yoga during one of the most tumultuous and inspired periods of his life, doing so long before he issued his warning about the dangers of kundalini.

The Swiss psychiatrist (1875–1961) and founder of analytical psychology turned to the discipline relatively early in his career as he struggled with two crises. The first was personal. In his thirties, as part of his inquiries, Jung engaged in a furious battle to pry open his own mind, so much so that he would often shudder with hallucinations and cling to nearby objects to keep from falling apart. Ultimately, his "confrontation with the unconscious," as he called it, resulted in a secret journal bound in red leather that, when published in 2009, was hailed as the genesis of the Jungian method.

The other crisis was World War I. It raged beyond the psychoanalyst and his home in neutral Switzerland, shattering the old European order. Jung perceived an enigmatic link between the inner and outer conflicts. And, in the interest of science, he used that relationship to push himself to what he considered the edge of madness. "I was frequently so wrought up," Jung recalled, "that I had to do certain yoga exercises to hold my emotions in check." He did so sparingly. "I would do these exercises only until I had calmed myself enough to resume my work with the unconscious."

Another example of the ostensible interaction between yoga and creativity centers on Leopold Stokowski (1882–1977), a conductor renowned for his exuberance, intuition, and a style that shunned the traditional baton for hand motions. He is often remembered for his starring role with the Philadelphia Orchestra in the Disney film *Fantasia*.

Early in his career, Stokowski became a confirmed health enthusiast, throwing himself into a disciplined regimen of yoga, meditation, and strict limits on what he ate and drank. He was said to be able to relax completely at will and, on six hours of sleep, handle workdays running up to eighteen hours. Before each concert, he would meditate to clear his mind.

Stokowski was also a famous womanizer. When, in the 1930s, he and Greta Garbo (1905–1990) found they could, so to speak, make beautiful music together, they traveled to Italy and, in the ancient town of Ravello, rented a villa overlooking the Mediterranean. There he taught her yoga. The actress, in turn, adopted the discipline wholeheartedly, studying with such teachers as Devi—famous as the first yoga teacher to the stars.

Garbo became such a devoted fan that she not only spread the word among friends and acquaintances but even played the teacher. Gayelord Hauser, a health guru of the day who advised the actress on dietary matters, recounted how Garbo taught him to do the Headstand. He found it rejuvenating. But Hauser also learned that it could damage the neck. Ultimately, he recommended avoiding the pose in favor of relaxing on a slanted board that lowered the head and raised the feet.

The world of classical music provided another possible example of how yoga can foster the creative impulse. Yehudi Menuhin (1916–1999) was a prominent violinist and conductor. Born in New York City, he performed hundreds of times for Allied troops during World War II and, as the soldiers liberated the German concentration camps, for the inmates who managed to survive. Many were little more than skeletons. In 1947, in a courageous act of reconciliation, he traveled to Berlin and became the first Jewish musician to perform in Germany following the Holocaust.

During this period, the exhaustions of conflict as well as the unstructured nature of his early training conspired to cause Menuhin great physical and artistic hardship. By the early 1950s, he was complaining of aches and pains, of tension and deep fatigue, of the impossibility of getting any rest. His art suffered.

Then, in 1952, while visiting India, he met Iyengar. The yogi taught him how to relax in Savasana, the Corpse pose. The musician immediately fell into a deep sleep. The ensuing yoga lessons gave Menuhin feelings of deep refreshment, as well as better control of his violin. Menuhin became a huge fan. In 1954, he gave Iyengar an Omega watch engraved on the back: "To my best violin teacher." Soon, the musician was introducing Iyengar to audiences in Britain, France, and Switzerland. It was Menuhin who put the unknown yogi on the world stage.

In 1965, when *Light on Yoga* came out, Menuhin wrote a foreword of considerable grace and passion. The star of classical music praised the discipline as giving a new perspective "on our own body, our first instrument," teaching individuals how to draw out the "maximum resonance and harmony." And Menuhin, a witness to war, recommended yoga as a path to virtue.

"What is the alternative?" he asked. "Thwarted, warped people condemning the order of things, cripples criticizing the upright, autocrats slumped in expectant coronary attitudes, the tragic spectacle of people working out their own imbalance and frustration on others." By nature, Menuhin concluded, yoga cultivated a respect for life, truth, and patience. He saw its civilizing qualities as implicit "in the drawing of a quiet breath, in calmness of mind and firmness of will."

More recently, the rock star Sting (who plays not only guitar but the lute) has praised yoga. He told an interviewer that it can produce a state of inner calm in which music comes to him as if from another dimension. "I don't think you write songs. They come through you," he said. "Yoga is just a different route to that same process."

What inspires such artists as Sting and Menuhin, Stokowski and Garbo, Jung and many other innovative minds, is impossible to know, as is precisely how yoga may have influenced their careers. Still, the question is worth asking given the discipline's deep resonance not only with celebrated artists but a variety of modern practitioners as well.

A cottage industry has sprung up in recent years that employs yoga as a means of inspiration. Yoga as muse gets promoted in workshops, books, retreats, travel tours, classes, and magazine articles, as well as by coaches and consultants. It is a little-known but increasingly common testament to yoga as a path to artistry.

"Yoga won't make writing easy," says Jeff Davis, a teacher, "because, well, writing is difficult. But yoga is helping thousands of writers to facilitate and design their own creative process—rather than to be at the whim of random flashes of inspiration, moods, or energy peaks."

Linda Novick is a painter who calls the Berkshires home but likes to travel to Miami Beach in the winter, Tuscany in the spring, and back to the Berkshires for the summer and fall. She also teaches yoga, and uses it to inspire her painting students. Her website, www.yogapaint.com, advertises her classes and philosophy. "Let go of fear and blocks to creativity," it counsels. Novick's book, *The Painting Path*, outlines gentle yoga exercises and uplifting thoughts that culminate in art projects, including ones in pastels, watercolor, batik, collage, and oil painting.

Mia Olson, a flautist, was teaching at the Berklee College of Music in Boston when she fell in love with yoga. She signed up for a teacher training course at Kripalu and began sharing yoga tips with her Berklee peers. Soon, she offered a class, Musician's Yoga, and was quickly asked to open another section. "The students," she recalled, "were craving this connection with mind and body."

The inspirational power of yoga seems to arise—at least in part—from nothing more complicated than the release of psychological tension and the quieting of the mind. Over the ages, many artists have looked to quiet for insight, exhibiting what Emily Dickinson called an "appetite for silence." The quietude let them see things differently.

That yoga can produce this state seems beyond doubt. In metabolic terms, the quieting depends on physiological cooling and the kind of relaxation response that Benson documented. Experience shows, however, that the path can be rocky.

Most yoga teachers, and many practitioners, know how a seemingly dull routine can erupt in sudden displays of upheaval. Mel Robin, in one of his books, called it not unusual for a beginning student toward the end of class to break down into "muffled sobs and copious tears." He suggested that yoga's lessening of tension can result in bursts of long-suppressed emotion.

Over the decades, several kinds of popular psychotherapy have sought to use physical leverage as a way of releasing and neutralizing toxic emotions. The methods include Rolfing, Neo-Reichian massage, Holotropic Breathwork, and Somatic Psychology. All seek to undo body tension as a way of breaking through mental blockages.

A few studies have shown that yoga can unlock the unconscious and liberate not only long-buried emotions but other feelings and thoughts, images and memories. While the general phenomenon is well known, the creative implications are seldom explored.

Elmer Green, a psychologist who studied Swami Rama, proved to be an exception. At the Menninger Foundation in Kansas, he and his wife, Alyce, examined the roots of creative reverie in college students. The scientists trained the students in biofeedback as well as the methods of Swami Rama, including rhythmic breathing and progressive muscle relaxation of the kind done in Savasana. The main part of the experiment focused on college juniors and seniors from Washburn University in nearby Topeka. The students did their calming routines in a dimly lit room and then sat back in a reclining chair while the scientists measured their brain waves and tape-recorded their answers to questions. On their own, the students also practiced the relaxation methods on school days for about an hour, and came back to the laboratory once every two weeks for the recording and interview sessions. In all, twenty-six students took part in the study.

The scientists reported that the exercises promoted "a deeply internalized state" that resulted in a range of insights and beneficial moods.

One student told of how he had gathered material for a paper but then got worried and tense after the flu interrupted his studies and left him feeling like he had lost focus and momentum. The problem, he reported, felt "insurmountable." Then a session left him very relaxed and his mind drifted through all the material. Suddenly, "everything just seemed to fall together."

The Greens proposed that the benefits spoke to a universal mechanism. If the students had been mature scientists, they argued, their insights might have centered on mathematical or chemical problems. Instead, the students found that the relaxation led to better relationships, greater concentration, more confidence, enhanced skills at organizing materials, and, in general, improvements in handling life challenges.

Artists, the Greens concluded, have no monopoly on imaginative solutions. The problems of living are "also amenable to insight, intuition, and creativity."

The cottage industry employs similar methods. In his book on writing, Davis recommends postures and types of breathing and awareness meant to quiet the cerebral din and help writers come up with fresh ideas. Doing something as simple as inhaling as long as one exhales, he advises, can become "a quick way to calm the chatter."

Science, it turns out, has uncovered at least one biochemical factor that promotes the quieting. It is GABA, the neurotransmitter we visited in the chapter on moods. Remarkably, its calming action has much in common with a much more famous way that artists have slowed their minds in order to aid their explorations.

Faulkner, Hemingway, Capote, and many other writers and artists in the twentieth century found not only comfort but inspiration in the bottle. The inebriation was so ubiquitous that a book, *Hemingway & Bailey's Bartending Guide to Great American Writers*, details the favorite drinks that the literary set imbibed in pursuit of relaxation and compositional fire.

Alcohol is a depressant that works beautifully to slow the brain. But its side effects are nasty. In the body, ethyl alcohol breaks down into toxins that can promote cancer as well as liver and brain damage, among other troubles.

Yoga is kinder. Yet its ability to calm the mind—to produce "a retardation of mental functions," as Behanan put it—shares a common biochemical basis with alcohol. Both do at least part of their mental rejiggering by means of GABA, or gamma-aminobutyric acid. The neurotransmitter slows the firing of neurons, making them less excitable and thus calming the mind. Ethyl alcohol does the trick indirectly. Its binding to neurons produces a chemical environment that increases the power of the inhibitory neurotransmitter.

By contrast, yoga's action is direct. The Boston team found that doing yoga caused levels of the potent neurotransmitter to rise, in one case nearly doubling. As many a yoga practitioner can attest, one result is a sense of physical and mental calming, of increased relaxation and reduced anxiety. Perhaps it is also the stuff of poetic inspiration.

• • •

Another factor in the quieting of the mind centers on the differing nature of the brain's hemispheres. In everyday life, the left side dominates. It excels at logic and language, as well as the din of cerebral chatter. But an emerging body of scientific evidence suggests that yoga can activate the brain's right hemisphere—the one that tends to govern intuition, creativity, instincts, aesthetics, spatial reasoning, and the sensing and expressing of emotion. So the discipline may act as an inspirational force in part because it shifts the hemispheric balance toward a more artistic frame of mind.

It has taken decades for scientists to tease the secrets of hemispheric character into the open and learn how the brain's two halves deal with the world in remarkably different ways. When the Greens did their studies in the 1960s and 1970s, the details were sketchy. But soon thereafter, the field made rapid progress, thanks in no small part to the investigations of Roger Sperry, a neurobiologist at the University of Chicago and the California Institute of Technology. In 1981, he won a Nobel Prize for his trouble.

Sperry focused on epileptics who had undergone an operation to ease their seizures. It severed the corpus callosum—the bundle of nerves that transmits signals between the brain's right and left hemispheres. Sperry and colleagues gave these patients special tasks. The surprising results showed that the differing sides of their brains had distinctive forms of consciousness. In effect, Sperry showed that every individual on the planet is endowed with not one but two brains, each pursuing its own particular way of thinking, perceiving, remembering, reasoning, willing, and emoting. His discoveries threw a generation of neuroscientists into uncovering the details of hemispheric specialization.

Today, the most basic difference between the two halves is considered to be how they process information. The right brain (which controls the body's left side) does its handiwork in parallel fashion—taking in many streams of information simultaneously from the senses and creating an overall impression of smell and sound, appearance and texture, feeling and sensation. For instance, the right brain dominates an inconspicuous type of sensory activity that yoga seeks to develop—proprioception, or inner knowledge of limb position. On the mat or in life, it tells us the position of our arms and legs—even with eyes shut. Proprioception, like other body functions dominated by the right brain, works best at portraying the big picture, at delivering impressions. It produces what is known in psychology as a gestalt, where the whole is greater than the sum of the parts. It is holistic.

By contrast, the left brain works in a sequential fashion. It excels at logic and language, math and science, reading and writing. The left brain revels in detail, in pattern recognition, in making judgments of social rank, and in putting things into the order of past, present, and future.

The right brain could hardly be more different. It is timeless and nonverbal, dealing in the eternal now, in the universe of sensory experience and emotion. It sees a flower and rejoices at its beauty and wholeness. The left brain sees the differing parts—the stem and petals, stamens and pistils. It anticipates the steps needed to bring the flower indoors—the shears, the vase, the water, the display setting. The right brain sees the flower as a lover would, the left brain as a florist.

The right brain's lack of regimentation makes a mess of exacting requirements but stands out when it comes to creativity, to seeing things in new ways, to thinking outside the box. It explores the possibilities of the moment. It cares little for social judgments but revels in spontaneity and adventure. It sees chaos not as a misfortune but as an opportunity for fresh perceptions and novel insights. It celebrates the new.

Modern neuroscience holds that many aspects of creativity (like most complex tasks) require the contributions of both halves of the brain and their complementary skills. An example is learning to play music. The left brain excels at such narrow responsibilities as reading notes, memorizing an instrument's pattern of fingering, and drilling scales over and over. The right brain adds the zing. It introduces the spice of improvisation, of playing by ear, of endowing the score with the color of emotion and personal interpretation.

Jill Bolte Taylor, a brain scientist trained at Harvard, has detailed many of these findings in her remarkable book *My Stroke of Insight*. Dramatically, she told of how she suffered a left-brain stroke that turned the abstractions of hemispheric differentiation into a riveting drama. A blood clot the size of a golf ball destroyed her powers of analysis and language, leaving her stranded in the joyous, peaceful, intuitive, sensory-rich world of her right brain.

"I felt like a genie liberated from its bottle," she wrote. "The energy of my spirit seemed to flow like a great whale gliding through a sea of silent euphoria." Luckily for happy endings and the writing of books, Taylor in time recovered her left-hemisphere skills. That success led her to feel a consuming urge to tell not only about the surprising plasticity of the human brain but the benefits of learning how to empower its right side.

Taylor portrayed the first step of the rightward shift as a willingness to live in the moment, in the here and now. The mind has to slow down, to lessen the left brain's fixation on analyzing and deliberating. She recommended drawing attention to breathing, to relaxing, to focusing on the constant stream of sensory information, and to feeling the resulting sensations. Her advice resembled the Buddhist practice known as mindfulness as well as the kind of awareness that yoga recommends, especially in Savasana. It also recalled the kind of relaxation that the Greens had the college students do.

In passing, Taylor mentioned yoga as a way in which many people "shift their minds." But she gave no details of how yoga works and limited her remarks to general observations about paying attention as a means of shifting the balance of hemispheric dominance.

Recent science has suggested that yoga and meditation can in fact stimulate the workings of the right brain. The studies tend to be small and preliminary but are nonetheless intriguing. Andrew Newberg, a doctor at the University of Pennsylvania Medical Center in Philadelphia, led much of the research. In the 1990s, he began to study whether experienced meditators could alter the workings of their brains. In 2001, he and his colleagues reported that brain scans of eight meditators showed an increased flow of blood in the right thalamus. The pair of small organs above the brain stem and below the corpus callosum relay sensory messages to the outer brain and the hypothalamus—the control center of the autonomic nervous system and the body's metabolic pitch.

Newberg and colleagues presented a more detailed portrait in 2007, drawing on results with twelve meditators and a control group. Here, too, the scientists found increased activity in the right thalamus.

Yoga eventually caught Newberg's eye. He did a preliminary study that involved two men and two women, their mean age forty-five. None of the subjects had significant experience in yoga or meditation, and all underwent three months of Iyengar training. The subjects performed their yoga routines daily, initially with a teacher and eventually at home with a DVD. The routine ran for roughly an hour and consisted of more than a dozen poses, including the Downward Facing Dog (Adho Mukha Svanasana) and the Seated Forward Bend with Bent Leg (Janu Sirsasana). The students also did rhythmic breathing in the form of Ujjayi pranayama as well as progressive relaxation and meditation.

Seated Forward Bend with Bent Leg, *Janu Sirsasana*

The scientists scanned the brains of the subjects at the start of the three months and at the end. In 2009, Newberg and six colleagues reported the results. "We found greater overall activations in the right hemisphere rather than the left," the scientists wrote. The areas of heightened blood flow included the frontal lobe, the seat of higher consciousness, and the prefrontal cortex, the well-developed region of the brain that distinguishes humans from other mammals. Both areas are important to setting and achieving goals, such as accomplishing the precise limb rearrangements of Iyengar yoga.

In closing, the investigators added that scientists in the future would have to conduct more thorough studies to sharpen their understanding and discover which parts of the typical yoga routine most influenced the rightward shift.

Over the decades, science has identified another aspect of hemispheric specialization that appears to bear strongly on the issue of creativity as well as the artistic lifestyle—if such a thing exists. The evidence suggests that the right hemisphere orchestrates not only emotion and spatial reasoning but the primal rumblings of sex.

The clues emerged as neuroscientists moved from electroencephalograms to scans that let them see heightened activity in the deep recesses of the brain. The studies linked sexual excitement to the lighting up of the right hemisphere and in particular its frontal and prefrontal areas. In seeking to explain the findings, scientists proposed that the frontal regions of the brain were producing the racy images and thoughts basic to sexual arousal—the glitter of daydream and desire, memory and fantasy. The studies indicated that the frontal regions tended to retain their glow even as levels of sexual excitement rose and (in step with fast breathing and other physical accelerations) the brain shifted its overall emphasis from cortical to limbic control.

While the association of sex and artistry may be new to neuroscience, it is extremely old stuff for the world at large and long predates Freud's theories about sexual energy being a stimulus to creativity. Indeed, the portraits of artists so regularly depict them as beholden to Eros that the image of promiscuity is a literary cliché. The list of the famously profligate includes not only Garbo and Stokowski but Oscar Wilde, Modigliani, Dylan Thomas, Jack Kerouac, Goya, Picasso, Marlon Brando, Hemingway, Frida Kahlo, and hundreds of others. The free spirits are seen as embracing whatever comes their way, be it lovers, food, or intellectual passions.

Science has addressed the issue and found evidence that supports the risqué stereotype. In 2006, the *Proceedings of the Royal Society of London*, one of the world's most venerable journals, reported on a study of four hundred and twenty-five British men and women. The inquiry categorized the levels of creativity among the subjects into four groupings—none, hobby, serious, and professional. The scientists found that the serious artists and poets on average had twice as many sex partners as the less creative types. Moreover, the professional artists tended to have the most lovers of all.

What all this means for yoga is unclear. The complexities of the brain and behavior are legion, as are the difficulties of establishing cause and effect. But yoga's ability to promote a rightward shift would seem to reinforce the idea that the discipline can act as a sexual tonic. At a minimum, the finding adds to the existing evidence about yoga's stimulating effects on human sexuality, as we saw in the case of hormones and brain waves. And it may ultimately shed light on human behavior. For the moment, the rightward shift suggests what might be considered a possible clue to how the discipline goes about heightening the artistic impulse.

The connections between sex and creativity become most evident with the kundalites. Their declarations of inspired artistry, coupled with new candor about the role of sexuality, seem to offer, at least in theory, an intriguing augmentation to Freud's ideas about the role of sexual energy. If Freud was right about creativity, and if the yogis are right about the inner fire surging into a sexual blaze, then perhaps kundalini does in fact provide a basis for artistic expression.

One way to investigate the issue is to see if any creative parallels to the kundalini experience have arisen and found their way into the deliberations of science. As it turns out, serious investigators have studied whole classes of individuals whose personalities have undergone sudden transformations.

An astonishing case involves Tony Cicoria, a former college football player who became an orthopedic surgeon. One fall afternoon in 1994, Cicoria was at a family gathering in upstate New York when he stepped outside a lake pavilion to call his mother. He was forty-two and in excellent health. The day was pleasant. But Cicoria, while approaching a pay phone, noticed dark clouds on the horizon. As he talked, it began to rain. He heard distant thunder. Cicoria had hung up and was about to head back to the pavilion when lightning flashed out of the phone and struck him in the face.

He fell to the ground. Sure he was dead, he saw people running toward his body, saw his children and felt they would be okay, saw the high and low points of his life. Waves of bliss and bluish-white light washed over him as he felt his consciousness starting to race upward. "This is the most glorious feeling I have ever had," he began to think. And at that instant—*bam!* He was back in his body.

Cicoria survived. Indeed, he soon found himself fit enough to resume work as a surgeon and once again move ahead with his life. But he was a changed man—a deeply changed man.

Within weeks, a longing for classical music replaced his love of rock. He acquired a piano and taught himself how to play. Soon, his head filled with music from nowhere. Within three months of the lightning strike,

Cicoria had little spare time for anything but playing and composing. Eventually, his marriage fell apart. But Cicoria pressed ahead. In 2007, he started giving recitals. In 2008, the Catskill Conservatory sponsored his debut at the Goodrich Theater in Oneonta, New York, where he lives. The sold-out audience was all smiles and applause. Also that year, Cicoria issued a CD of classical piano solos titled *Notes from an Accidental Pianist and Composer*. Prominent among the arrangements was "The Lightning Sonata."

Oliver Sacks, the distinguished author and neurologist at Columbia University, details the case of Cicoria in his fascinating book *Musicophilia*. He also discusses other examples of people who have experienced a sudden passion for art and music. Sacks cites a body of developing evidence that traces such transformations to traumatic rewirings of the brain, in particular its limbic system and its temporal lobes, home of the hippocampus and long-term memory as well as auditory processing.

The surges appear similar to what happens to kundalites. If Cicoria experienced a blinding flash from outside his body, the kundalites seem to experience a similar shock from within. Indeed, some yogic authorities liken the mystic current to a bolt of lightning.

So does kundalini stir creativity? No scientific studies have addressed the issue. But the anecdotal evidence is rich.

Gopi Krishna (1903–1984), the Kashmiri who inspired Pond and his friends, reported that the stabilization of his own inner fire coincided with the commencement of an unending flow of poetry. The pandit composed verse in not only his native Kashmiri but Urdu, Punjabi, Sanskrit, Persian, Arabic, French, Italian, English, and German. It was an urge he was unable to extinguish.

Krishna—who went to college for two years in Lahore but failed the examination that would have let him continue his studies—claimed to have little or no knowledge of several of these languages. Instead, he said the poetry welled up from inside him, as if from a universal source. At times, his mind rebelled when his inner voice told him that a poem was about to emerge in a foreign tongue.

"I had never learned German," he recalled protesting at one point, "nor seen a book written in the language, nor to the best of my knowledge ever heard it spoken."

Carl von Weizsäcker (1912–2007), an eminent German physicist whose brother served as president of West Germany, wrote the introduction to Krishna's book *The Biological Basis of Religion and Genius*. There he said that he found the German poetry to be rustic but inspired, much like a folk song. "It is, if one may say so, touching," he wrote. He gave a few sample lines as well as a translation:

Ein schöner Vogel immer singt
In meinem Herz mit leisem Ton

A beautiful bird always sings
In my heart with a soft voice

"What makes this poetic phenomenon possible and what purpose does it serve?" von Weizsäcker asked. "I do not know. Honor the incomprehensible!" From someone else, such a proposition might have sounded irresponsible. But the German physicist had discovered such basic things as how big stars like the sun generate their energy.

Untold numbers of kundalites have undergone artistic makeovers similar to Krishna's. Franklin Jones, a California guru who in the 1980s moved to Fiji, produced a diverse body of artwork ranging from cartoons to ink brush paintings to giant works to multiple-exposure photographs, including many studies of the female nude. His 2007 book, *The Spectra Suites*, showcased some of the results. By the time he died in November 2008, his oeuvre ran to more than one hundred thousand works.

Jana Dixon, a kundalite I visited in Boulder, argued that her own inner fire had inspired her artwork. Her *Biology of Kundalini* website has a page devoted to her paintings, and I saw canvases in various states of completion around her apartment. Her images were electric in color and design, some bordering on the psychedelic, some unabashedly erotic.

"When my K is up," Dixon told me, "it's peak creativity."

It was in Canada that I found the most ambitious studies of kundalini and creativity—the core objective of the Institute for Consciousness Research. If cultivation of the mystic fire represents a dangerous undertaking, as Jung warned, the group's investigations seemed to suggest that kundalini also has a primal upside.

• • •

The kundalites looked quite unmystical—some frumpy, some turning gray, some thin and elegant, all seemingly part of the upper middle class and glad to be chatting with one another in rural Ontario on a summer weekend. They wore sandals and shorts, baggy pants and flowery shirts, running shoes and cotton frocks. All had plastic name tags. The group seemed about evenly divided between men and women. They sat attentively in a big white tent filled with fifty or sixty plastic lawn chairs and listened to speakers recount some very personal experiences, the presenter occasionally pausing in tense silence, head down, holding back tears. They took long breaks for schmoozing and eating—lots of eating. The meals featured lush vegetarian dishes and salads dotted with blueberries. Big cookies appeared at coffee breaks.

"We're everyday people," Dale Pond, one of the organizers, told me during a break, her voice slightly edgy. "We do wine and cheese parties." Indeed, every night, Paul and Dale Pond invited the kundalites over to their house a few miles down the road to party, Ontario style, with good beer and snacks.

It was the late summer of 2009 and the occasion was the twenty-fourth annual conference of the Institute for Consciousness Research. The group's original name captured its early affability: Friends in New Directions, or FIND. The conference site was a farm about two hours north of Toronto. The spot was beautiful and private. Thick stands of conifers surrounded the old barn, the farmhouse, and the wide lawn that held the big tent. Just off the main highway, to mark the turnoff, a temporary sign had been set up that pointed down a long gravel road. "FIND-ICR," it said, welcoming friends old and new. Although the group's core members remained in Ontario, attendees came from such places as Baltimore and San Francisco, New York and Pennsylvania. Not all were kundalites. But all had developed an interest in the subject and, most especially, its creative repercussions.

For this annual meeting, the organizers put the focus on the personal stories, as suggested by the conference title: "Kundalini: Changing Lives from Within." The speakers told of how the mystic fire had touched them and displayed the results in the form of songs and poems, meditations and paintings.

The informal agenda seemed just as important. A table displayed kundalini books that were for sale, including nearly a dozen by Krishna.

Perhaps most important, the relaxed atmosphere gave time for networking and comparing notes. It was a quiet place where people could talk about their experiences, their coping strategies, their dreams.

Teri Degler, a writer who had profiled several of the assembled kundalites in her books, and who had undergone her own ecstasy of arousal, joked about how the word "kundalini" could be loosely rendered as their own peculiar brand of craziness: "Kind of Loonies."

A businessman told me how much he enjoyed the get-togethers and how he found it impossible to speak of his kundalini experience at work. "What would I say? 'Hey, wait a second, guys. I've got a wind blowing up my back.'"

Paul Pond, a lean man of sixty-three, opened the program and ran it like a veteran. He joked a lot and had a deadpan style that kept the audience in high spirits. But his introductory tour of the kundalini horizon was dead serious. He touched on all the major issues—the sexual nature of the experience, the joys, the dangers, and the subtle repercussions. Standing at a white podium under the billowing tent, speaking into a microphone, Pond said kundalini awakenings seemed to be on the rise and that the wave could prove important in stabilizing the wobbly planet. "We need direction," he said, "and that's going to come from within."

Pond said historical researchers had shown that kundalini arousal tended to foster the creative fires and complimented the speakers for agreeing to speak frankly about their own experiences and struggles.

His wife, Dale, described her own. She had been profiled by Degler in a book, *Fiery Muse*. It said Dale had been a shy woman who lacked a serious intellect when, two decades ago, she underwent a kundalini arousal that transformed her into a serious reader, a productive artist, and confident public speaker.

At the podium, she reiterated those claims. "I did spontaneous art, spontaneous poetry," Dale told the audience of her early days. "All the different parts of me were opening up." The inner fire, she said, fostered a deep sense of inner cohesion and inspiration that—like the musical compositions of Tony Cicoria and the poetry of Gopi Krishna—seemed to come from nowhere. "I'd be crying, watching myself do art, and say, 'Where did *that* come from?'"

Under the tent, speaker after speaker struck related themes. Neil Sinclair—the chairman of CyberTran International, a start-up in Richmond, California, that is seeking to create a highly ecological passenger railroad—stepped to the podium in sandals, white socks, and a flowered shirt. He told of how kundalini had struck in 1973 while he was a freshman at the University of California at Berkeley. The setting was a Halloween party. Sinclair had dabbled in yoga and meditation for many years. During the party, he retreated to an empty bed as his mind began to reel. He felt a release at the base of his spine followed by an upward sense of expansion.

"It didn't stop," he told the audience. "A rush came up and I lost any sense of my body and I found myself immersed in an expanding sphere of ecstasy." He called it "an orgasmic sensation" that seemed to engulf the universe.

Sinclair cautioned the uninitiated to avoid thinking of kundalini as unmitigated bliss. "Gopi Krishna almost died twice," he noted. "He was on the verge of insanity. Society is not there cheering you on. It's very challenging."

He peppered his talk with readings from the poetry he began to write shortly after his awakening. He said the words tended to tumble into his head.

A book of Sinclair's poetry had just been published, titled *The Spirit Flies Free: The Kundalini Poems*. During a break, I bought a copy. It contained more than a hundred poems whose topics ranged from war and apple trees to the workings of the harpsichord. Several struck wilderness themes. Mystic reflections ran throughout the volume. But Sinclair kept the fundamentals simple, as with the opening lines of the collection:

Beneath the surface of this world,
Invisible to the naked eye,
Exists an energetic framework,
The basis of both you and I.

Over the years, a number of intriguing clues about the relationship between yoga and creativity have come to light. It seems like they now constitute a significant body of evidence. Still, the findings are relatively modest. Other topics more central to the discipline—health, fitness, safety—have received more attention.

One reason for the comparatively slow advance is sheer complexity.

By definition, creativity goes to deep issues of psychology and ultimately what it means to be human—areas that science has always had a hard time investigating. Science tends to do the easiest things first. It is nothing if not practical. This fact of scientific life suggests the magnitude of the challenge that investigators face.

Even so, the importance of the subject and the potential richness of the returns make it attractive. Big risks can produce big rewards. It is the kind of topic that might flourish in the decades ahead.

The cottage industry might grow into schools. Maybe cures would emerge for creative paralysis. Creative blocks might go extinct. Perhaps many people would learn how, as Menuhin put it so eloquently, to draw out their "maximum resonance." Maybe world leaders would take up yoga as an aid to their deliberations, formalizing the kind of reflective calm that Larry Payne introduced at Davos.

Maybe yoga would soar.

Epilogue

Run the clock forward a century or two. What is yoga like? It seems to me that, based on current trends, two very different outcomes are possible. Both revolve around science, otherwise known as the pursuit of systematized truth.

In one scenario, the fog has thickened as competing groups and corporations vie for market share among the bewildered. The chains offer their styles while spiritual groups offer theirs, with experts from the various camps clashing over differing claims. Immortality is said to be in the offing. The disputes resemble the old disagreements of religion. But factionalism has soared. Whereas yoga in the late twentieth century began to splinter into scores of brands—all claiming unique and often contradictory virtues—now there are hundreds. Yet, for all the activity, yoga makes only a small contribution to global health care because most of the claims go unproven in the court of medical science. The general public sees yoga mainly as a cult that corporations seek to exploit.

In the other scenario, yoga has gone mainstream and plays an important role in society. A comprehensive program of scientific study early in the twenty-first century produced a strong consensus on where yoga fails and where it succeeds. Colleges of yoga science now abound. Yoga doctors are accepted members of the establishment, their natural therapies often considered gentler and more reliable than pills. Yoga classes are taught by certified instructors whose training is as rigorous as that of physical therapists. Yoga retreats foster art and innovation, conflict resolution and serious negotiating. Meanwhile, the International Association of Yoga Centenarians is lobbying for an extensive program of research on new ways of improving the quality of life among the extremely old. Its president, Sting, recently embarked on a world tour to build political support for the initiative.

In short, I see the discipline as having arrived at a turning point. It has reached not only a critical mass of practitioners but a critical juncture in its development.

Yoga can grow up or remain an infant—a dangerous infant with a thing for handguns. Traditionalists may find it loathsome. But growing up in this case means that yoga has to come into closer alignment with science, accelerating the process begun by Gune, Iyengar, and the other pioneers. The timeless image is a mirage. Yoga has changed many times over the centuries and needs to change again.

The stakes are enormous—and not just for the millions of practitioners who expect a safe experience. The really gargantuan issue is helping the discipline realize its potential.

I caught a glimpse of the future that Friday night at Kripalu when Amy Weintraub said, "It really saved my life." Her testimony still rings in my ears, giving me hope for better ways of fighting the blues.

In antiquity, the geniuses of India forged a radically new kind of relationship between humans and their bodies. We are now on the cusp of learning how to apply their discoveries in startling new ways, of bestowing on the world new gifts of healing and emotional renewal, health and vitality, personal energy and creative inspiration. Think of Loren Fishman holding up his healed arm. Think of Amy Weintraub doing Breath of Joy. Physicians talk about breakthroughs in personalized medicine and pharmacogenetics—of using information from a person's genetic map to tailor medicine to his or her own particular needs. But yoga can already do that. It can turn our bodies into customized pharmaceutical plants that churn out tailored hormones and nerve impulses that heal, cure, raise moods, lower cholesterol, induce sleep, and do a million other things. Moreover, yoga can do it at an extremely low cost with little or no risk of side effects. It has the potential to usher in a genuine new age, not one of wishful thinking.

Western science tends to view the body as a fixed thing with unchanging components and functions. But yoga starts from a different premise. It sees a lump of clay. The body in this view is awaiting the application of skilled hands.

A conviction of some Hindus and spiritual yogis is that we live in the Kali Yuga—a dark time in which people are distant from God and civilization has fallen into decline. They venerate the past. With all due respect, I see the best times for yoga as lying ahead. We can turn the fledgling discipline into a better shaper of clay.

If yoga played for keeps, if it achieved a new kind of maturity, the discipline could become a force in addressing the global crisis in health care, which in the United States now consumes more than $2 *trillion* a year. It could become the basis for an inexpensive new world of health care and disease prevention, of healing and disciplined well-being. It might be a game changer. Michelle Obama is working hard to achieve those kinds of benefits for young people.

But to have a hope of exerting greater influence on the organization of global health care, yoga must come into closer alignment with science—with clinical trials and professional accreditation, with governmental authorities and their detailed evaluations, probably even with insurance companies and their dreaded red tape. Yoga could become a major force. Or it could stay on the sidelines, a marginal pursuit, lost in myths, looking to the past, prone to guru worship, fracturing into ever more lineages, increasingly isolated as the world moves on.

Realizing even a small fraction of yoga's potential is going to require work—hard work.

We need to make advances along two complementary lines of inquiry that, as this book demonstrates, have coexisted since the start of the scientific investigation of the practice: We must better understand what yoga can do and better understand what yoga can be. The latter issue goes to Robin's "better yoga."

Let's call the postural discipline that yogis started practicing in medieval times Yoga 1.0. The modern variety that formed early in the twentieth century under the influence of science might be called Yoga 2.0. Now Yoga 2.5 or even 3.0 seems to be in the works, judging from the advent of many vigorous styles and the wide efforts of yoga professionals to make their discipline safer. In the future, Yoga 4.0 may yet emerge, quite different from anything we can now imagine.

A first step in yoga's wider development centers on addressing the threat that practitioners face right now—the lack of reliable information about the discipline's pros and cons. Increasingly, it seems like the din of competing styles, the rise of new commercial ventures, and the inchoate nature of Yoga 3.0 are adding to the confusion. I have tried my best to clarify the situation with this book (and its suggestions for further reading and detailed notes). But there's still a long way to go—and a lot more that can be done—to help make trustworthy information more widely available.

One problem is the diffuse nature of the existing science. It seems fairly unique in having been done in so many places over such a long period of time. In my travels, I was impressed at how experts had assembled troves of books and papers. The Ponds in Canada, Sat Bir Khalsa in Boston, Mel Robin in Pennsylvania, Gune's ashram south of Bombay, and PubMed in Bethesda have all assembled much good information on the science of yoga. But they all seem to have different pieces of the puzzle. And I suspect there are many more out there waiting to be uncovered, examined, and shaped into a comprehensive body of knowledge.

If I could snap my fingers and make it happen, I would establish a Yoga Education Society that took on the job of pulling all the information together and making it publicly available. YES could become not only a central repository but an impartial voice that summarized the information, giving practitioners a good place to go for reliable assessments. YES could also act as a force to counteract the growing waves of commercial spin and help raise the visibility of yoga benefits that seem to get relatively little attention, such as the discipline's promise as an antidepressant, a sex therapy, and a stimulus to creativity.

If I have been hard on yoga commercialization, it is because the trend raises fundamental questions that seldom get addressed. Today, as always, yoga has no social mechanism that sifts through the numerous claims to ascertain the truth, and the commercial blitz with its dynamic goals and competitive agenda seems to make that weakness all the more glaring. Imagine if Big Pharma had no Food and Drug Administration and other regulatory agencies looking over its shoulder. The marketing of fake diseases and bogus cures—already a multibillion-dollar embarrassment despite all the bureaucratic scrutiny—would be much worse.

Yoga seems to be moving toward that kind of predatory behavior as it grows into a bustling industry. Of course, commercial ventures can also perform wonderful acts of public service. Witness the free event with all the yogis in Central Park. But what they do best is promote their own interests and welfare.

To me, the great hope of improvement centers on expansions of scientific research and the rise of the kinds of thoughtful individuals profiled in this book. They are busy combining yoga and science, leaving behind the ambivalence of recent decades and looking ahead. The group represents a vanguard of forward thinkers with serious degrees, serious interests, and—perhaps most important—the serious credibility required to raise the discipline's standing. They are changing both what yoga is and our understanding of what it can do.

The decades between the founding of Gune's ashram and the publication of *Light on Yoga* bore witness to a radical shift of perspective. Yoga, instead of looking to gurus and antiquity for guidance, looked to science. But that bond weakened over the years. As a result, yoga's primal attitudes often reasserted themselves.

Today, it seems that the relationship between science and yoga is ripe for revitalization. I take heart not only from the new generation of scientific yogis but from the declarations of respected authorities such as the Dalai Lama, the Tibetan spiritual leader. In his book *The Universe in a Single Atom*, he writes that "spirituality must be tempered by the insights and discoveries of science." Remarkably, he even states that if science found particular tenets of Buddhism to be false, "then we must accept the findings of science and abandon those claims."

Another encouraging sign is that government authorities in the United States and elsewhere have started to fund the science of yoga, mainly as a means of evaluating the discipline's potential for disease prevention and treatment. The goal is to document the true benefits. In Bethesda, Maryland, the National Institutes of Health, the world's premier organization for health-care research, is spending money and raising standards. It began funding yoga research in 1998 and has now paid for dozens of studies, including investigations of yoga's ability to treat arthritis, insomnia, diabetes, depression, fatigue, and chronic pain. Many of these studies appeared since I began my inquiry in 2006, suggesting that the pace of scientific research is quickening. The wave tends to be high quality, helping raise yoga's social credibility.

These public investments are starting to pay off in terms of treatments and insights, as suggested by some of the most interesting reports in this book. The Institutes funded the hypertension study in Pennsylvania, the cardiovascular study in Virginia, the telomere study in California, the aerobics study in New York, the neurotransmitter study in Boston, the right-brain study in Philadelphia, and the musician study in Massachusetts, among other projects. Such inquiries are revealing true paths to a better future.

In 2011, the Institutes began a new cycle of studies, despite increasingly tight budgets. They include yoga for cancer survivors, for adults who suffer persistent depression, and for elderly women at risk of cardiovascular disease.

Opponents of federal research love to disparage yoga investigations as extravagant wastes of taxpayer money. In 2005, *Human Events,* a conservative journal, ridiculed yoga studies as symptomatic of the "bloated bureaucracy syndrome." Such criticism is likely to grow in the years ahead as political battles heat up in Washington over how to reduce the federal budget deficit.

It follows that the public funding of yoga research, without concerted advocacy, is unlikely to see significant increases anytime soon. Wherever you live—in the United States or elsewhere—it seems like a good time to write your representatives or take other steps to bring the merits of yoga studies to the attention of public officials. In 2011, the amount of money that the National Institutes of Health spent on yoga research amounted to about $7 million. That's too small to qualify as even a drop in Washington's bucket. It's nearly invisible. A much larger investment seems wise, given that yoga's demonstrated skills at disease prevention might result in savings of billions of dollars in traditional health-care costs. The outlay is highly leveraged, as actuaries like to say.

As a society, we are learning that extended old age can mean extended pain and debilitation, with worn-out organs and crippling dementias turning the twilight years into tragedies. Yoga seems to hold out the promise of increasing not only our life spans but our health spans. It may be part of the answer to enhancing not just the quantity of life but its quality, to helping us remain healthy for a longer period of time, to making our last years more vital and productive. That promise seems like a wonderful topic for a serious program of research.

The stakes go far beyond practicalities. One of the most interesting frontiers has little or nothing to do with expediency and everything to do with simple understanding.

What if Paul had been able to do a few brain scans and other measurements while the Punjab yogi sat in his deathlike trance? What new science might have emerged? Is disanimate bliss a human birthright? Is the euphoric trance safe? Can it spiral into madness? Does it make you a better person? Can it improve how we treat one another?

Science fiction with its portrayals of long space fights that feature coffinlike freezers and frozen astronauts may be passé. Perhaps human hibernation—as Paul described it more than a century and a half ago—is the right way to go. Maybe future astronauts will slip into a Full Lotus when voyaging between the stars.

We have yet to address scientifically—much less begin to unravel—such questions. At a minimum, a deeper understanding of yoga has humanitarian implications ranging from practical therapies for people caught in kundalini's coils to psychoanalytic insights of a kind that Jung would have cherished.

The public evidence suggests that yoga's rather profound ability to slow the human metabolism can function like a match to ignite a sexual blaze. Often, the resulting state is feverish and the yogi animated (if not meditating or immobilized in the Punjab yogi's kind of catalepsy). As noted in chapter 6, I call this reversal the yoga paradox. It has received no explicit attention to my knowledge from either yoga professionals or the world of biomedicine. Corby's team at Stanford saw glimmers of the transformation. The main symptom is a radical change of homeostasis—the body's metabolic equilibrium—from cool to hot. One of my hopes for this book is that it will prompt the scientific community to carefully study this and other aspects of yogic hypersexuality.

The science of yoga has only just begun. In my judgment, the topic has such depth and resonance that the voyage of discovery will go on for centuries, perhaps millennia. What started with Paul and studies of respiratory physiology will spread to investigations ever more central to life and living, to questions of insight and ecstasy, of being and consciousness. Ultimately, the social understanding that follows in the wake of scientific discovery will address issues of human evolution and what we decide to become as a species.

Even so, as I mentioned in the prologue, it's important to remember that science has no monopoly on the truth.

As a science journalist, I have devoted my career to writing about science and trying to illuminate its findings and methods. Science is incredibly tough in practice despite its often gentle and glamorous image. By nature, it seeks to limit the role of faith, to make as few assumptions as possible, and to subject the information it gathers as well as its own tentative findings to withering doubt. A synonym for "science" is "organized skepticism." The process can be intellectually brutal. The constructive side is that science, done right, also works to suspend judgment, to collect and test and verify before coming to firm conclusions. In theory, it can see without prejudice. That makes it a rare thing in the world of human institutions.

But science—even at its best, even with its remarkable powers of discrimination and discovery—is nonetheless extraordinarily crude. It ignores much about reality to zero in on those aspects of nature that it can quantify and comprehend. What gets set aside can be considerable—the wonders of the Sistine Chapel, among other achievements. Science, for all its triumphs over the last four centuries, sometimes fails to see the obvious. It is blind to the individuality of a snowflake and the convulsions of the stock market, not to mention ethics. No equation is going to outdo Shakespeare.

My book *The Oracle* devoted its last chapter to sketching out the limitations of scientific knowledge. The arguments are philosophic in nature but come down to the great difficulty that science faces in trying to provide a comprehensive worldview.

What I know with certainty is that science cannot address, much less answer, many of the most interesting questions in life. It's one finger of a hand, as a wise man once said. I treasure the scientific method for its insights and discoveries, as well as for the wealth of comforts and social advances it has given us. But I question the value of scientism—the belief that science has authority over all other interpretations of life, including the philosophic and spiritual, moral and humanistic.

So while the science of yoga may be demonstrably true—while its findings may be revelatory and may show popular declarations to be false or misleading—the field by nature fails utterly at producing a complete story. Many of yoga's truths surely go beyond the truths of science.

Yoga may see further, and its advanced practitioners, for all I know, may frolic in fields of consciousness and spirituality of which science knows nothing. Or maybe it's all delusional nonsense. I have no idea.

But even if the otherworldly view has merit, this book and the long studies of the scientific community show the bottom line. The transcendental bliss starts with the firing of neurons and neurotransmitters, with surges of hormones and brain waves.

It's the science of yoga.

Afterword: The Outcry

My own evaluation sees the rewards as exceeding the risks, a conclusion *The Science of Yoga* supports in hundreds of pages and studies. The benefits are many and commonplace while the serious dangers tend to be few and comparatively rare. I know this happy outcome from practicing yoga since 1970 and from researching the book for five years. Even so, its publication stirred more controversy than anything else I've ever written as a science journalist—mainly, it seems, because the book tells of hidden dangers that can cripple and kill.

Yoga for many people is a sacred refuge. It inspires feelings of trust and loyalty that few social institutions now enjoy. For many practitioners, the ties go beyond family and community. Identifying the risks of anything so revered is likely to upset. But as I learned of the dangers, I felt an obligation to help people disentangle the good from the bad, improve their practice, and become their own best teachers. After all, yoga too often gets sold as completely safe—"as safe as mother's milk," as a prominent guru once declared (see page 103).

Despite the controversy, or perhaps because of it, the book and the uproar appear to have done much to improve yoga around the world. I'm enormously proud of having taken on what turns out to have been a topic of lengthy and secretive deliberation—proud even when the controversy has turned personally abusive. I sleep well knowing that the book has, in all likelihood, helped people avoid a number of incidents that would have ended in disabilities and perhaps even death.

Before publication, Bobby Clennell, my talented illustrator and a senior Iyengar teacher with a global following, told me that the book would start a conversation. The reality was more like a riot.

The outcry began when an excerpt ran in *The New York Times Magazine* in January 2012 and ricocheted around the globe under the headline "How Yoga Can Wreck Your Body." A furious yogi shot back: "*You are a wreck.*" It probably didn't help that the magazine's editors chose to illustrate the excerpt with photos of the Broadway cast of *Godspell* twisted into exaggerated poses. The postures were meant to be funny. But lots of yogis took offense. As a *Chicago Tribune* columnist remarked, the article provoked "more coverage, umbrage and yuppie outrage than the wars in Afghanistan and Iraq combined."

Readers had no way of knowing that the excerpt had been drawn from just one of seven chapters and that the overall book discussed not only dangers but a wealth of positive news on such topics as health and healing, sex and longevity, moods and creativity. Most readers had no idea that I'm a yoga booster, not a basher. As I like to tell audiences, I did yoga this morning and will, God willing, do it tomorrow. But that message got lost amid the cyber outrage.

Many hundreds of emails, letters, and communications have come my way since the excerpt's appearance in January 2012 and the book's publication in early February. (*The Times Magazine*, before pulling the plug, registered 737 comments.) I'm happy to report that the *vast* majority of the feedback has been overwhelmingly positive. Expressions of gratitude have come from yoga teachers and celebrities, doctors and therapists, yoga schools and studio owners—in general, from people of long experience. Many told of injuries. A teacher of nearly two decades wrote with elegant simplicity: "Thank you."

To my surprise, I learned that a number of accomplished yogis had written whole books about yoga injuries and how to avoid them. The books got little attention when published but demonstrated the depth of concern. Kevin Khalili, a doctor in Santa Barbara, California, wrote *X-Posed*, published in 2011. Jean-Paul Bouteloup, a yoga teacher in Paris, wrote *Yoga sans dégâts* (Yoga Without Damage), published in 2006. And the genre turned out to be relatively old. In 1980, an early student of B.K.S. Iyengar—one of modern yoga's founders—issued a remarkable book in Paris: *Attention, le yoga peut etre dangereux pour vous!* (Warning—Yoga Can Be Dangerous for You!). The author, Noelle Perez-Christiaens, coauthored the work with Louis Creyx, a doctor. The publisher was Institut de Yoga B.K.S. Iyengar. So their book was no exposé but rather a sharp warning from insiders.

Some letters moved me deeply because people told in considerable detail of upheavals in their lives. The damage included strokes, spinal stenosis, nerve injury, disk rupture, and dead spinal tissue. "I am currently recovering from cervical fusion and will need a lifetime of physical therapy," a former studio owner wrote. An injured teacher told of her ordeal in a twenty-eight-line poem.

One of the saddest and most thoughtful letters came from an elderly man who had studied with Iyengar in India for sixteen years. His list of personal injuries included torn ligaments, damaged vertebrae, slipped disks, deformed knees, and ruptured blood vessels in his brain. "All that you wrote," he said, "I can confirm in my own life." He concluded his long letter by saying he was almost done writing a book about his experiences.

I had been accused of sensationalism. But the spate of letters argued otherwise and prompted me to track down the latest federal statistics in an effort to better understand the big picture. The book noted (page 121) that a sample of hospital emergency rooms in 2002 had reported a sharp rise in yoga injuries to forty-six. That turned out to be a preview. By 2011, the annual sample had more than doubled to one hundred cases. The increase seemed to reflect yoga's growing popularity as well as practitioners doing more aggressive styles. What I had failed to realize in writing the book was that Washington, eager to find and reduce life dangers, routinely used these ER reports to compile *national* injury estimates. It extrapolated—carefully. Its statisticians had selected roughly a hundred emergency rooms whose locations and demographics mirrored the nation's 5,300 hospitals with emergency services, and the monitoring of that subset led to the statistical portraits. The result was, in effect, an early warning system.

For yoga, the numbers rose, as the chart on the next page reveals. In the book, I suggested (page 122) that "many hundreds or even thousands" of cases went unreported. My guess turned out to be low. Over the past decade, the federal estimates of national harm add up to more than 20,000 injuries. By definition, that figure excludes yogis who went to medical clinics, family doctors, yoga therapists, urgent care centers, and other healers—or who kept their injuries to themselves. Federal officials told me that the national estimates speak exclusively to emergency rooms. So the *really* big picture is unknown.

Federal Estimates of Injured Yogis Admitted to Hospital Emergency Rooms

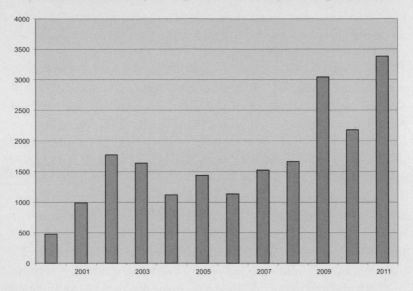

Source: *National Electronic Injury Surveillance System
of the Consumer Product Safety Commission*

If twenty million people in the United States have practiced yoga in the past decade, the 20,000 figure means the fraction of practitioners who went to emergency rooms was just 0.1 percent—or 1 person in 1,000. That's a low rate of injury. Critics have noted that the rates for activities such as weight lifting and golf are higher. Fair enough. All exercise involves some degree of risk, and yoga is no different (though countless gurus and teachers have claimed otherwise). The trouble with rate comparisons is that they focus exclusively on the *quantity* of injuries and pay no attention to their *quality*, or severity. But a sprained ankle is obviously very different from a stroke. No study that I know of has made *qualitative* comparisons between the injuries of sports and yoga. Perhaps they are similar. After all, the damage is typically minor—strains and sprains. But the hospital reports demonstrate that yoga can also result in fracture, seizure, dislocation, rupture, heart failure, spinal displacement, deep vein thrombosis, vertigo, nerve injury, and brain damage. In time, research will likely clarify the differences. It may turn out that yoga, despite a low rate of injury, is unusually dangerous.

Today—given my tour of yoga's inner sanctums—I could write an encyclopedia of dysfunction. Yet I still see the rewards as outweighing the risks. The proviso is that yoga has to be done intelligently. The solution is smarter yoga, better yoga, safer yoga. I dropped the Plow and the Shoulder Stand when I learned that doctors had linked these poses to arterial damage that can keep blood from the brain and turn agile practitioners into cripples (see pages 111 to 121).

The Science of Yoga seems to be improving safety around the globe. Over and over, I've heard of classes discussing the pros and the cons of different styles and poses. "How NOT to Wreck Your Body," read the ad for a Boston class. "We will examine the 'most dangerous' yoga postures and offer safe methods of practice." Many people have told me of old routines updated, of new precautions taken, of fresh emphases given to fitting the pose to the person rather than twisting bodies and physical idiosyncrasies into highly idealized postures. Significantly, a number of yoga schools now use the book in teacher training.

"Thank you for opening up a door," wrote a woman, a studio owner for nine years. The new candor, she added, promised to usher in "improved safety for students—and a positive yoga experience."

On Sunday, April 15, 2012, I participated in a panel discussion on the topic of injuries at a *Yoga Journal* conference in New York City. The magazine's editor, Kaitlin Quistgaard, presided. It was four against one—but I knew what I was getting into and had decided that it was important to take the issue to yoga gatherings whenever possible. The main arguments were that yoga, like any other sport or life activity, has risks, and that the postures are intrinsically safe but can be misapplied in ways that harm.

All that seemed fairly reasonable, even though the points were somewhat contradictory. What I found surprising was the strong defensiveness about yoga strokes—to me, the worst of the dangers, and one of the best documented. The pushback seemed aimed at denying that the problem existed at all.

Frankly, that troubled me. *Yoga Journal*—before Quistgaard joined the magazine—had written about yoga strokes and how the brain damage could result in crippling disability and even death, as the book notes (see pages 126 to 128). Was it now backpedaling? Was Quistgaard feigning disbelief? Had the outcry so polarized the yoga community that its leaders felt obliged to strike an attitude of denial? Or—perhaps worst of all—was Quistgaard simply unaware of her own magazine's reporting? In our discussion, I was especially concerned by what I considered a lack of understanding about the strength of the scientific evidence.

During the initial outcry, yogis had expressed similar doubts and I had addressed their questions in a number of media interviews. Because strokes can result in lengthy disability and even death—because the stakes are so extraordinarily high—I have gathered the most common questions below and have attempted to answer them, making what I consider a strong case for informed caution. I also quote some of the stroke victims. To me, their warnings are the most compelling of all.

You cite a 1972 letter to the editor of the British Medical Journal—*nothing more than correspondence. You call that evidence?*

A number of top journals classify their most important reports as "Letters" or, in some cases, "Correspondence." The reason is that scientific journals began centuries ago with the aim of distributing the letters of scientists—and they still reflect that origin, especially British journals. In 1951, when Watson and Crick announced their discovery of DNA's structure, they did so in a "letter" to *Nature*, a prestigious British journal. As for yoga strokes, the physician who wrote the 1972 letter to the *British Medical Journal* was one of England's most distinguished neuroscientists, and his warning proved to be visionary.

The reports of strokes you cite are extremely small in number and incredibly old—the first in 1973. Get real! The risk is obviously exaggerated because there are no modern reports!

The excerpt in *The Times Magazine* amounted to a sketch of the book's disclosures, so its readers saw relatively little evidence concerning the issue's persistence to the present day. But there is much. First, the book detailed how three early reports got amplified in warnings and supplementary findings, as noted on pages 121 (of the text) and 251 (of the endnotes). In particular, it cited medical reviews in 1989, 1994, and 2001 that described yoga strokes, as well as a recent book that did likewise. In addition, it cited a 2009 survey of yogis. That investigation (discussed on page 134) reported four strokes—four modern cases in which yogis, not the medical establishment, told of brain damage. That showed plainly that the danger was not an aberration of the past but a reality of the present. The newest evidence comes from the federal monitoring of emergency rooms. The network's report of 2011 tells of a man who was attending a yoga class when he suffered a stroke. He was rushed from the emergency room into the hospital.

Despite such public and private reports, a skeptic might ask why medical journals—arguably the most important way that doctors share information—present no fresh cases of people who have suffered yoga strokes. The answer, at least in part, seems to center on how the medical establishment perceives the danger as well known. Science by nature is highly progressive. Journal editors want fresh ideas, not confirmations of old ones. No editor is now accepting manuscripts on new proofs of DNA's helical structure. That finding is old and accepted—like yoga strokes.

I am appalled at your omission of comparative studies. All kinds of activities result in strokes. Why not compare the risks?

I too am appalled. Yoga is a huge enterprise and I wish the medical community devoted more time and energy to its study. Unfortunately, no scientist has published data on how often yogis suffer strokes and compared the rate with other sources of brain damage. I think it's fair to assume that the risks are low—but how low is an open question. The book (on pages 114 to 115) details another kind of activity that can trigger the strokes. Known as the beauty-parlor syndrome, it happens when women during shampooing have their necks tipped too far back over the edge of a sink. Here too, unfortunately, the true incidence is unknown.

So where are the bodies? You claim that strokes kill. But yoga authorities say they know of no deaths from yoga injuries—none, ever—for any reason at all, much less strokes. Why are you so alarmist?

I wish the facts were otherwise. But the public record shows that many yogis in apparent good health die suddenly—at times during hard training. Jeff Goodman, fifty-eight, died during advanced instruction in Houston in April 2012. Tiffany Neff, twenty-five, died a month earlier in Moreland Hills, Ohio. Jules Paxton, forty-five, passed away in New York City in 2011, as did Bill Jackson, fifty-five, in Naples, Florida.

Eric Berliner, fifty-eight, died in Chicago in 2010 during a master class. Abbey Duncan, twenty-seven, a yoga teacher, died that same year in Minneapolis, Minnesota. The list goes on and on. In Los Angeles in 2004, Sita White, forty-three, a British heiress and a favorite of gossip columnists, collapsed and died in a yoga class. The question is why.

Medical authorities in some cases cite natural causes. Does that rule out yoga? Jeff Goodman, the Houston yogi, was said to have died of a brain aneurysm—the thinning and ballooning of an arterial wall. Burst aneurysms have been linked to stationary exercises such as push-ups and weight lifting, which produce spikes in blood pressure as the breath is held and the abdominal muscles are contracted. Yogis do that all the time. It happens in poses that lift substantial body weight and in exercises that bottle up the breath. So too, headstands send blood rushing downward and can raise pressure in the head—so much so that the faces of students can become flushed and their eyes bloodshot. Can yoga rupture blood vessels in the brain?

The emergency room reports tell of a woman who was taking a yoga class when she suddenly felt a "ripping sensation" in her brain and passed out. She was admitted to a hospital and diagnosed as having suffered an intracranial hemorrhage. So too, the man who studied with Iyengar told me of blacking out for an hour after a vigorous session and having a physician identify the underlying problem as a cerebral hemorrhage. In both cases, the brain ruptures never got written up in the scientific literature and never received notice in the public communications of medicine, as best I can determine. Searches for the hemorrhages in electronic databases came up empty. Instead, these two cases of yoga causing damage to the body's most important organ were known intimately to only a few people and represent the typical situation for the countless diagnoses that working doctors make every day.

Yoga authorities are right in saying that injuries have resulted in no reported deaths. But that speaks to the limits of data collection and reporting in biomedicine rather than a demonstrable absence of fatalities among millions of yoga practitioners. It's a state of ignorance versus one of knowledge. Yoga is a mysterious universe. Science has explored aspects of it (many of which I discuss in the book), but most of its worlds remain unknown and uncharted, at least in terms of modern surveillance and experimentation.

Even *if* the understandings of basic science seemed to indicate a credible link between yoga and a particular fatality, a practical impediment would probably get in the way. Crime shows tend to portray forensics as a glamorous field. But a recent study by the National Academy of Sciences reveals the discipline of postmortem inquiry to be crippled by old rules, obsolete equipment, scarce funds, and inadequate staffing. Remarkably, the study told of how one jurisdiction recently conferred the title of coroner on a high school senior. The report called autopsies—which cost about $2,000—"largely budget driven." That means that government cutbacks are turning some classes of death investigation into unaffordable luxuries.

The institutional failings and the knowledge gaps make determining the underlying cause of death especially difficult in the case of stroke. The most common type begins when the blood supply to the brain fails. The challenge is pinpointing the blockage.

The investigation requires imaging equipment of considerable sophistication and a thorough study of the welter of arteries that loop through the head, neck, and upper chest. The early doctors conducted just that kind of study in their original linking of yoga and brain damage. But the challenge can be so great that—for a large number of strokes—physicians cannot find where the obstruction of the blood supply began. The medical term for such brain damage is *cryptogenic*, meaning its origin remains a mystery. Fatal strokes present an added barrier to understanding because doctors obviously cannot ask the deceased about any activity that may have been the underlying cause of death—be it yoga or something else. In contrast to this uncertainty, identifying the source of a cerebral hemorrhage can be as easy for a doctor as peering into a skull and seeing a ruptured blood vessel, perhaps a burst aneurysm.

Given the difficulties, the issue of how many people die from yoga strokes quickly becomes an exercise in estimation. *Yoga Journal*—in the article that predated Quistgaard's arrival at the magazine—examined the main kind of stroke associated with yoga and cited a medical report that put the number of fatalities at something less than one person in twenty. That figure can lead to extrapolations of the overall death rate. But that's statistical, not observational. In the years ahead, it seems likely that the medical community will become more attuned to the threat and do a better job at uncovering its incidence.

Why does this issue upset you so? It seems that you have become even more agitated since the book's publication.

It's the letters—letters from yogis who have written to me about their strokes and their upended lives. I feel entrusted with information that I am under a moral obligation to share. During public talks, I often read the letters aloud. They have prompted me to do what I can to alert yogis to what is probably an extremely low risk but one that can have extremely high consequences—if not in sending practitioners to the grave, then in crippling them and diminishing their lives.

Here are two excerpts, both quoted with the permission of the authors. The first is from a yoga beginner, a man. At 2:30 in the morning, a few hours after a yoga session, he suddenly found himself reeling physically as the part of his brain that regulates fine muscular activity, the cerebellum, began to malfunction.

"This left me on the bathroom floor thrashing around in a pool of my own vomit crying for help," he wrote. "I was fortunately rushed to the hospital. The end result was lots of physical therapy to learn to balance and walk again, double vision, and some loss of hearing in my left ear. Prior to this, I had gone to a few yoga classes but never once was I aware that practicing yoga could cause serious injuries. Although I feel that I have made a good recovery, this incident has ruined my life both physically and emotionally. Thank you for making everyone aware of the dangers."

The second letter is also from a man, a practitioner with more experience. His doctors linked his stroke to the Plow.

"I remember thinking as I was doing the exercise that it felt like an unhealthy amount of pressure on my neck," he wrote. "After the yoga series, I had a crushing headache, which was a rarity for me. The next morning, I noticed my left eyelid was drooping. At first, I thought it was possibly due to a sinus infection, because I had a cold at the time. But later I noticed that my left pupil was constricted. It took weeks of trips to different doctors before I was diagnosed." His doctors told him that the odds of surviving such an event are about one in ten thousand. The man had surgery to correct his drooping eyelid as well as four years of drugs, therapy, and brain scans. "It's good," he wrote, "that your article presents a warning."

Such letters keep me doing what I can to explore the science and alert the yoga community to the known risks. I keep seeing the letters, keep thinking about the man's odds of survival being put at one in ten thousand. That's infinitesimal. The federal government cites the same number as the likelihood that an individual will, over a lifetime, get hit by a bolt of lightening.

It seems that the right way to honor the injured is to learn, to pay attention to their experiences, and to see what the damage reveals about the dangers and their possible reduction or elimination. Yoga is too great an enterprise to harbor inadvertent failings. My hope is that their stories will—like the book and the outcry—help stimulate discussion of ways to better limit the risks and multiply the rewards. Anything less would dishonor the constructive spirit of yoga and neglect what seems like a genuine opportunity to do some good.

William J. Broad
October 2012

Further Reading

Here are some recommended books on the science and history of yoga, as well as a few selections from related fields. The list makes no claim of being comprehensive but simply offers entree to a growing literature that draws on demonstrable fact and reasonable inference to illuminate yoga.

Michael J. Alter. *Science of Flexibility*, 3rd ed. Champaign, IL.: Human Kinetics, 2004. *The inside story on extreme bending.*

Loren Fishman and Carol Ardman. *Relief Is in the Stretch: End Back Pain Through Yoga.* New York: Norton, 2005. *A guide and rationale.*

Loren Fishman and Ellen Saltonstall. *Yoga for Arthritis.* New York: Norton, 2008. *A strategy and how it works.*

Judith Hanson Lasater. *Yogabody: Anatomy, Kinesiology, and Asana.* Berkeley: Rodmell Press, 2009. *A tour of the inner body for better practice and teaching.*

William D. McArdle, Frank I. Katch, and Victor L. Katch. *Exercise Physiology: Nutrition, Energy, and Human Performance,* 7th ed. Philadelphia: Lippincott Williams & Wilkins, 2009. *A bible of sports science that features hundreds of informative graphics.*

Timothy McCall. *Yoga as Medicine: The Yogic Prescription for Health and Healing.* New York: Bantam, 2007. *A thoughtful guide rooted in science and personal observation.*

Mel Robin. *A Physiological Handbook for Teachers of Yogasana.* Tucson: Fenestra Books, 2002. *A classic—only 629 pages long.*

_____. *A Handbook for Yogasana Teachers: The Incorporation of Neuroscience, Physiology, and Anatomy into the Practice.* Tucson: Wheatmark, 2009. *The updated classic—only 1,106 pages!*

Robert M. Sapolsky. *Why Zebras Don't Get Ulcers*, 3rd ed. New York: Henry Holt, 2004. *A lucid examination of how prolonged stress can result in major afflictions.*

Richard M. Schwartzstein and Michael J. Parker. *Respiratory Physiology: A Clinical Approach.* Philadelphia: Lippincott Williams & Wilkins, 2005. *The scientific basics from physicians at the Harvard Medical School.*

Mark Singleton. *Yoga Body: The Origins of Modern Posture Practice.* New York: Oxford University Press, 2010. *How Hindu nationalism and early health fads helped create modern yoga.*

Hugh B. Urban. *Tantra: Sex, Secrecy, Politics, and Power in the Study of Religion.* Berkeley: University of California Press, 2003. *Light on the world of yoga eroticism.*

Amy Weintraub. *Yoga for Depression: A Compassionate Guide to Relieve Suffering Through Yoga.* New York: Broadway Books, 2004. *Beautifully written advice on mood lifting and how it works.*

David Gordon White. *Kiss of the Yogini: "Tantric Sex" in its South Asian Contexts.* Chicago: University of Chicago Press, 2006. *High scholarship on early sexual rites and centuries of misrepresentations.*

Notes

Chronology

xxv *earliest known precursors of yoga:* Thomas McEvilley, "An Archaeology of Yoga," *Anthropology and Aesthetics,* no. 1 (Spring 1981), pp. 44–77; Gregory L. Possehl, *The Indus Civilization* (Lanham, MD: AltaMira Press, 2003), pp. 141–45. See also Mircea Eliade, *Yoga: Immortality and Freedom* (Princeton, NJ: Princeton University Press, 1990), pp. 353–358. Note that recent scholarship has cast doubt on this traditional interpretation. See, for instance, Geoffrey Samuel, *The Origins of Yoga and Tantra: Indic Religions to the Thirteenth Century* (New York: Cambridge University Press, 2008), pp. 2–8, and David Gordon White, *Sinister Yogis* (Chicago: University of Chicago Press, 2009), pp. 48–59.

xxv *citation as a founding document:* Mark Singleton, *Yoga Body: The Origins of Modern Posture Practice* (New York: Oxford University Press, 2010), pp. 26–27. The date for the *Yoga Sutras* comes from White, *Sinister Yogis,* p. xii.

xxv *Erotic sculptures of the Lakshmana temple:* David Gordon White, *Kiss of the Yogini: "Tantric Sex" in its South Asian Contexts* (Chicago: University of Chicago Press, 2006), pp. 98, 144–46.

xxv *Gorakhnath, a Hindu ascetic:* David Gordon White, *The Alchemical Body: Siddha Traditions in Medieval India* (Chicago: University of Chicago Press, 2007), pp. 90–101.

xxvi *describes a magic rite:* Gudrun Bühnemann, "The Six Rites of Magic," in David Gordon White, ed., *Tantra in Practice* (Princeton, NJ: Princeton University Press, 2000), p. 448.

xxvi *The Yoni Tantra:* Indra Sinha, *Tantra: The Cult of Ecstasy* (London: Hamlyn, 2000), pp. 135, 140–42.

xxix *begins spending public funds:* Gordon Edlin and Eric Golanty, *Health & Wellness,* 9th ed. (Boston: Jones and Bartlett, 2007), p. 434.

Prologue

1 *illustrating its parking tickets:* Laura Crimaldi and Ira Kantor, "Cambridge 'Yoga' Parking Tickets Have Drivers in a Twist," *Boston Herald*, www.boston herald.com, September 21, 2010.

2 *made it part of Let's Move:* Anonymous, "Take Action Kids: 5 Simple Steps to Success," www.letsmove.gov/sites/letsmove.gov/files/pdfs/TAKE_ACTION _KIDS.pdf.

2 *On the White House lawn:* Kaitlin Quistgaard, "Yoga Diary: Posing at the White House," *Yoga Journal* lifestyle blog, April 6, 2010, blogs.yogajour nal.com/yogadiary/2010/04/posing-at-the-white-house.html; Anonymous, "White House Yoga," New Image Photography, newimagephotography .com/blog/?p=1034.

2 *did a tricky balancing pose:* Anonymous, "Yoga at the White House, An Easter Tradition! 2011 Pics," www.yogadork.com/news/yoga-at-the-white -house-an-easter-tradition-2011-pics.

2 *puts the current number of practitioners:* Anonymous, "Yogamonth Media Kit 2010," Yoga Health Foundation, Venice, California, p. 11.

2 *a gathering of thousands:* Lizette Alvarez, "STRETCH; Yoga, Brought to You By . . . ," *New York Times*, June 27, 2010, Section MB, p. 7.

3 *yoga industrial complex:* John Friend—founder of the Anusara style and the company, Anusara, Inc.—is considered an exemplar of the commercialization trend. For a profile, see Mimi Swartz, "The Yoga Mogul," *New York Times Magazine*, July 25, 2010, Section MM, p. 38.

3 *charged small studios:* Nora Isaacs, "Hot, sweaty and scandalous," Salon .com, April 4, 2003.

3 *thousands of patents:* Suketa Mehta, "A Big Stretch," *New York Times*, May 7, 2007, Section A, p. 27.

3 *According to marketing studies:* Ronald D. Michman and Edward M. Mazze, *The Affluent Consumer: Marketing and Selling the Luxury Lifestyle* (Westport, CT: Praeger, 2006), p. 124.

3 *citing high incomes:* Anonymous, "The Growth of Yoga: Audience," *Yoga Journal*, 2008, www.yogajournal.com/advertise/pdf/YJ_audience_08.pdf.

4 *"cut with all kinds":* Bryant Urstadt, "Lust for Lulu: How the Yoga Brand LuLulemon Turned Fitness into a Spectator Sport," *New York*, August 3, 2009, p. 30.

4 *"The beginner":* I. K. Taimni, *The Science of Yoga* (Wheaton, IL: Quest, 1972), p. vii.

5 *"can drop to one-half":* Stanley W. Jacob and Clarice Ashworth Francone, *Structure and Function in Man,* 2nd ed. (Philadelphia: W. B. Saunders, 1970), p. 390.

6 *cargo-cult science:* Richard P. Feynman, "Cargo Cult Science: Some Remarks on Science, Pseudoscience, and Learning How Not to Fool Yourself," in Richard P. Feynman, *The Pleasure of Finding Things Out: The Best Short Works of Richard P. Feynman* (Cambridge, MA: Helix Books, 2000), pp. 205–16.

6 *It showed that scientists:* PubMed, www.pubmed.gov. I used "yoga" as a search term, doing so in 2006. As of 2011, the number of citations had grown to more than 1,600.

7 *able to download:* N. C. Paul, *A Treatise on the Yoga Philosophy* (Benares, India: Recorder Press, 1851), books.google.com/books?id=CZmNNpTy7VUC.

8 *two massive books:* Mel Robin, *A Physiological Handbook for Teachers of Yogasana* (Tucson: Fenestra Books, 2002); Mel Robin, *A Handbook for Yogasana Teachers: The Incorporation of Neuroscience, Physiology, and Anatomy into the Practice* (Tucson: Wheatmark, 2009).

9 *cripples more than one hundred million:* Anonymous, "Depression," World Health Organization, www.who.int/mental_health/management/depression/definition/en.

10 *prompting them to gain weight:* M. S. Chaya, A. V. Kurpad, H. R. Nagendra, et al., "The effect of long term combined yoga practice on the basal metabolic rate of healthy adults," *BMC Complementary and Alternative Medicine,* vol. 6, no. 28, published online August 31, 2006, www.biomedcentral.com/1472-6882/6/28.

10 *do fight pounds successfully:* Nicholas Bakalar, "Yoga May Help Minimize Weight Gain in Middle Age," *New York Times,* August 2, 2005, Section F, p. 7.

10 *As Carl Jung put it:* C. G. Jung, "Psychological Commentary," in W. Y. Evans-Wentz, ed., *The Tibetan Book of the Dead* (New York: Oxford University Press, 2000), p. xlvi.

I: Health

13 *an ugly little man:* Khushwant Singh, *Ranjit Singh: Maharajah of the Punjab, 1780–1839* (Hyderabad: Orient Longman, 1985), pp. 23–27; Surinder

Singh Johar, *The Secular Maharaja: A Biography of Maharaja Ranjit Singh* (Delhi: Manas, 1985), pp. 20–21.

13 *wandering yogi had approached:* William G. Osborne, *The Court and Camp of Runjeet Sing*, reprint of the 1840 edition (Karachi: Oxford University Press, 1973), pp. 123–38; James Braid, *Observations on Trance: or, Human Hibernation* (London: John Churchill, 1850), pp. iii–iv, 9–17; John Martin Honigberger, *Thirty-Five Years in the East* (London: H. Baillière, 1852), pp. 126–31; H. P. Blavatsky, "The Sadhoo's Burial Alive at Lahore: Important New Testimony," *The Theosophist*, vol. 2, no. 5 (February 1881), pp. 94–95.

13 *"a Hindoo idol":* quoted in Braid, *Observations*, p. 12.

14 *"for good compensation":* Richard Garbe, "On the Voluntary Trance of Indian Fakirs," *The Monist*, vol. 10, no. 4 (July 1900), p. 487; see also Osborne, *The Court*, pp. 170–71, and Braid, *Observations*, p. 20.

14 *he was presented with:* Braid, *Observations*, p. 14.

14 *They read palms:* George Weston Briggs, *Gorakhnath and the Kanphata Yogis*, reprint of the 1938 edition (Delhi: Motilal Banarsidass, 1989), pp. 1–25, 55; John Campbell Oman, *The Mystics, Ascetics, and Saints of India* (London: T. Fisher Unwin, 1903), pp. 168–86.

14 *To obtain new members:* Ibid., pp. 26–27; Singleton, *Yoga Body*, p. 117.

14–15 *prey on trade caravans:* White, *Sinister Yogis*, p. 201.

15 *"miscellaneous and disreputable vagrants":* Quoted in Briggs, *Gorakhnath*, p. 4.

15 *"homologous to the bliss":* White, *Kiss*, p. xxi. For more on the topic, see Samuel, *The Origins*, pp. 156, 283, 328.

15 *path to the ecstatic union:* Eliade, *Yoga*, pp. 49–50, 104, 200–273.

15 *under the pretext of spirituality:* Recent scholarship tends to associate the orgies with only a few sects and sees most Tantric lineages as practicing symbolic paths to blissful union. See, for instance, White, *Tantra in Practice*, pp. 4–5, 15–18. A counterpoint argument is that Tantra harbors a long tradition of public sanitization meant to hide its inner teachings. See Hugh B. Urban, *Tantra: Sex, Secrecy, Politics, and Power in the Study of Religion* (Berkeley: University of California Press, 2003), pp. 134–64. What seems undeniable is that many Tantric texts give explicit directions on how to perform acts of group and individual sex.

15 *eloping to the mountains:* Garbe, "On the Voluntary Trance," p. 499.

15 *nadir with the Aghori:* Eliade, *Yoga*, pp. 296–301.

16 *condemned as a threat to society:* Urban, *Tantra*, pp. 70–72.

16 *"Press the perineum":* Brian Dana Akers, trans., *The Hatha Yoga Pradipika* (Woodstock, NY: YogaVidya.com, 2002), p. 16.

16 *"a female partner":* Ibid., p. 72.

16 *"embraced by a passionate woman":* Ibid., p. 61.

16 *seldom refer to the origins:* While popular yoga usually avoids any reference to the Tantric roots of Hatha, scholars openly acknowledge the relationship. See, for instance, Eliade, *Yoga*, pp. 227–36, and James Mallinson, trans., *Gheranda Samhita* (Woodstock, NY: YogaVidya.com, 2004), p. xiv.

17 *Hatha means violence:* Monier Monier-Williams, *A Sanskrit–English Dictionary* (New Delhi: Asian Educational Services, 1999), p. 1287.

17 *a number of scholars:* See, for instance, White, *Kiss*, p. 217; Singleton, *Yoga Body*, p. 27; and Joseph S. Alter, *Yoga in Modern India: The Body Between Science and Philosophy* (Princeton, NJ: Princeton University Press, 2004), p. 24.

17 *The New Age approach:* A founder of modern yoga, B.K.S. Iyengar, helped establish this interpretation in popular culture. See his book, *Light on Yoga* (NY: Schocken, 1979), p. 439. Today, it is often the sole definition cited by yoga authorities. See, for instance, Martin Kirk and Brooke Boon, *Hatha Yoga Illustrated* (Champaign, IL: Human Kinetics, 2006), p. 2. Interestingly, Iyengar in his introduction to *Light on Yoga* speaks parenthetically of Hatha meaning "force or determined effort" but in his glossary goes further to define it as "against one's will." See pp. 22, 520.

17 *emphasis on the miraculous:* Eliade, *Yoga*, pp. 227–36, 274–84, 311–18. For a detailed account of yogis as sorcerers and miracle workers, see White, *Sinister Yogis*.

17 *scholar at Wesleyan University:* William R. Pinch, *Warrior Ascetics and Indian Empires* (New York: Cambridge University Press, 2006).

17 *"There was a clear tactical advantage":* Quoted in James Dao, "No Food for Thought: The Way of the Warrior," *New York Times*, May 17, 2009, Section WK, p. 3. For more on the warrior yogis, see Singleton, *Yoga Body*, pp. 39–40.

18 *"as if dead":* Akers, *The Hatha*, p. 111.

18 *"presumptuous to deny":* Quoted in Braid, *Observations*, p. 17.

18 *the birth of a new science:* For a sketch, see Singleton, *Yoga Body*, pp. 49–53.

18 *a passing reference:* Elizabeth De Michelis, *A History of Modern Yoga:*

Patanjali and Western Esotericism (London: Continuum, 2005), pp. 136–37; Singleton, *Yoga Body*, pp. 52–53.

18 *until I went to Calcutta:* My visit extended from Sunday, June 24, to Wednesday, June 27, 2007.

19 *the last volume:* Anonymous, *Report of the Late General Committee of Public Instruction for 1840–41 and for 1841–42* (Calcutta: National Library, Government of India, 1842), p. 105.

20 *heard lectures on such racy topics:* Anonymous, "List of Members," *Selection of Discourses Read at the Meetings of the Society for the Acquisition of General Knowledge* (Calcutta: Bishop's College Press, 1843), vol. 3, p. iv.

20 *his elite status:* Even so, Paul hailed from the very bottom of the social hierarchy. As a military doctor, he was a "Sub-Assistant Surgeon," a classification the colonial medical service held in low esteem. See Christian Hochmuth, "Patterns of Medical Culture in Colonial Bengal, 1835–1880," *Bulletin of the History of Medicine*, vol. 80, no. 1 (Spring 2006), pp. 39–72.

20 *seemed eager to show:* Paul, *A Treatise*, pp. iv, 52, 61; see also Helena Petrovna Blavatsky, *From the Caves and Jungles of Hindostan* (London: Theosophical Publishing Society, 1892), pp. 310, 315.

20 *such things as the Aghori:* Paul, *A Treatise*, p. 7.

20 *"abstaining from eating":* Ibid., p. iii.

20 *"has puzzled a great many":* Ibid., p. 43.

21 *It was carbon dioxide:* Following the science of his day, Paul consistently referred to carbon dioxide as carbonic acid. Here, I use the more familiar term. Paul's description was accurate because, in the environment, carbon dioxide quickly binds with water to form carbonic acid, which tends to be quite weak. Our moist breath is slightly acidic. In the atmosphere, rain and carbon dioxide mix to form what we call acid rain. Closer to home, carbonic acid is the secret ingredient that makes carbonated drinks such as soda and seltzer taste refreshingly tart.

21 *much about the basics:* Starting early in the nineteenth century, scientists tracked carbon dioxide out of unwarranted dread. Buildings were made to maximize ventilation and dilute stale air. People slept with their windows open, even in winter. See Jeff Stein, "How Things Work: An Interview with Michelle Addington," *Architecture Boston*, March–April 2005, pp. 44–49. The needless fear arose from misinterpretations of the experiments of Antoine Lavoisier, the father of modern chemistry. In the eighteenth century, he had put small animals under glass jars until they died. The

misapprehension arose that the animals had died from their own pernicious exhalations rather than lack of oxygen. In the first half of the nineteenth century, the forces of social progress united to fight the carbonic peril. The advertised dangers ranged from headaches to death. Interestingly, Paul turned this paranoia on its head, striking a blow for comprehension in an age of muddle. He showed that yogis experienced carbonic acid as a kind of elixir rather than a deadly poison.

21 *take fewer breaths:* Paul, *A Treatise*, pp. 8–11. He identified the practice by its correct name, *Kumbhaka*, which in Sanskrit means "like a pot" and connotes filling or holding. For a description, see B. K. S. Iyengar, *Light on Pranayama: The Yogic Art of Breathing* (New York: Crossroad Publishing, 2006), pp. 105–11.

21 *"one of the easiest methods":* Paul, *A Treatise*, p. 13.

21 *live in a gupha:* Yogic lore venerated such caves as holy ground. In Paul's day, sacred gupha dotted the Himalayas and other mountainous parts of India. Inevitably, as yoga grew in popularity, tourists began to seek out the caves. Today, package tours that focus on yoga and meditation often make stops at sacred gupha, after which exhausted sightseers make their way back to luxury hotels and restaurants.

21 *"a confined atmosphere":* Ibid., p. 3.

22 *drove his point home:* Ibid., pp. iv, 15–26, 36, 44.

22 *let the Punjab yogi survive:* Ibid., pp. 43–44.

22 *"promotes a build-up":* David A. Wharton, *Life at the Limits: Organisms in Extreme Environments* (Cambridge, England: Cambridge University Press, 2002), p. 77.

22 *began a revolution:* His 1851 book languished for years. Then, in 1880, *The Theosophist*, the monthly journal of the Theosophical Society, published in Bombay, began serializing Paul's book. Soon Indian presses were churning out new versions: a second edition in 1882 at Benares and in 1883 at Calcutta, a third edition in 1888 at Bombay, and a fourth edition in 1899 at Bombay. The printings spread Paul's naturalism across India and beyond.

23 *"as miraculous evidence":* Lee Siegel, *Net of Magic: Wonders and Deceptions in India* (Chicago: University of Chicago Press, 1991), p. 3.

23 *more than 6 percent:* B. D. Tripathi, *Sadhus of India: The Sociological View* (Bombay: Popular Prakashan, 1978), pp. 120–24, 217. For other accounts of burial tricks, see Siegel, *Net*, pp. 168–71. For doubts about the Punjab yogi, see Singleton, *Yoga Body*, p. 48.

23 *two holy men from India:* Richard Schmidt, *Fakire und Fakirtum im Alten und Modernen Indien* (Berlin: Verlag von Hermann Barsdorf, 1908), pp. 102–108; Garbe, "On the Voluntary Trance," pp. 481–82.

23 *famous for his precise studies:* Lucile E. Hoyme, "Physical Anthropology and Its Instruments: An Historical Study," *Southwestern Journal of Anthropology*, vol. 9, no. 4 (Winter 1953), pp. 408–30; Karl Pearson, "Craniological Notes: Professor Aurel von Török's Attack on the Arithmetical Mean," *Bibmetrika*, vol. 2, no. 3 (June 1903), pp. 339–45.

23 *in a preliminary report:* Aurel von Török, "Ueber die Yogis oder sog. Fakire in der Milleniums-Aasstellang zu Budapest," *Correspondenz-Blatt der deutschen Gesellschaft für Anthropologie, Ethnologie and Urgeschichte*, vol. 27, no. 6 (June 1896), pp. 49–50.

24 *sought to revive and modernize Hinduism:* Christophe Jaffrelot, ed., *Hindu Nationalism: A Reader* (Princeton, NJ: Princeton University Press, 2007), pp. 3–19. In the West, the image of Indian independence tends to focus on Gandhi and his campaign of nonviolence. But many Hindu nationalists called for violent struggle, and the political unrest resulted in a number of riots and killings. See Chetan Bhatt, *Hindu Nationalism: Origins, Ideologies and Modern Myths* (Oxford: Berg Publishers, 2001), pp. 78–79.

24 *symbols of all that had gone wrong:* Singleton, *Yoga Body*, pp. 4, 48, 117–18.

24 *"an embarrassing heritage":* Samuel, *The Origins*, p. 336.

24 *And it got one:* Urban, *Tantra*, pp. 134–64. Scholars have identified factors beyond science that aided yoga's modernization. For self-help methods of personal renewal, see De Michelis, *A History*, pp. 116–19. For physical culture, see Singleton, *Yoga Body*, pp. 81–162. For Hindu nationalism, see Joseph S. Alter, "Yoga and Physical Education: Swami Kuvalayananda's Nationalist Project," *Asian Medicine*, vol. 3, no. 1 (2007), pp. 20–36.

25 *first major experimental investigation:* Alter, *Yoga*, pp. 27, 30–31, 73–108.

25 *"sending out youths":* Anonymous (but clearly J. G. Gune), "The Kaivalyadhama: A Review of Its Activities from October 1924 to March 1930," *Yoga Mimansa*, vol. 4, no. 1 (July 1930), p. 75.

25 *"He never wanted":* Interview, O. P. Tiwari, secretary, Kaivalyadhama Yoga Ashram, Lonavla, India, June 27, 2007.

25 *threw himself into the nationalist struggle:* Here I follow the review of Gune's early life as recounted by Mandhar L. Gharote and Manmath M. Gharote, *Swami Kuvalayananda: A Pioneer of Scientific Yoga and Indian Physical Education* (Lonavla, India: The Lonavla Yoga Institute, 1999), pp.

11–22. Early in his career, Gune took the name Swami Kuvalayananda as a literary pseudonym. Since it was not a formal monastic title, I refer to him throughout this book as Gune.

26 *a wealthy industrialist:* Pratap Sheth was a rich Hindu nationalist and a whirlwind of philanthropy. In 1914, he founded the Khandesh Education Society, a private group that supported schools for Indian youth. In 1916, he founded the Indian Institute of Philosophy, which advocated yogic study. He also funded Balkrishna Shivram Moonje, one of the most militant early figures of Indian independence. To a remarkable degree, Sheth's agenda of putting social activism over asceticism prefigured the goals that came to characterize Gune's life as well as the reformulated yoga, making him a major if unknown figure in its rise. Pratap Sheth, sometimes written as Pratapseth, was also known as Agarwal Motilal. For a biographical sketch, see Anonymous, "Motilal Manekchand Agarwal," in Waman P. Kabadi, ed., *Indian Who's Who 1937–38* (Bombay: Yeshanand & Co., 1937), p. 479. For wider portraits, see Gharote and Gharote, *Kuvalayananda*, pp. 14–15, 24, 156, 158; G. R. Malkani, *A Life Sketch of Srimant Pratapseth: The Founder of the Indian Institute of Philosophy* (Amalner, India: Indian Institute of Philosophy, 1952). For his funding of Moonje, see Narayan Gopal Dixit, ed., *Dharmaveer Dr. B. S. Moonje Commemoration Volume: Birth Centenary Celebration, 1872–1972* (Nagpur, India: Centenary Celebration Committee, 1972), p. 74.

26 *benefactor again came to the rescue:* Gharote and Gharote, *Kuvalayananda*, p. 24.

26 *unique for the day:* A modern bibliography on the science of yoga lists Gune as the lead author on forty-eight papers—far more than any other investigator back then. See Trisha Lamb, *Psychophysiological Effects of Yoga,* International Association of Yoga Therapists, Prescott, Arizona, 2006), pp. 77–80.

26 *"We cannot make":* Anonymous (but clearly J. G. Gune), "Editorial Notes," *Yoga Mimansa,* vol. 3, nos. 3 and 4 (July–October 1928), second impression, 1931, p. 168.

26 *maintained a virtual taboo:* The word "Tantra" appears nowhere in the pages of *Yoga Mimansa* during the decades in which Gune ran the journal, according to a computer search of its texts that I performed in February 2010. Nor does the word appear in Gune's 1931 book, Swami Kuvalayananda, *Popular Yoga Asanas* (Rutland, VT: Charles E. Tuttle, 1974).

27 *found that the poses:* Anonymous (but clearly J. G. Gune), "Blood Pressure Experiments," "A Few More Figures of Blood Pressure," and "Yogic Poses and Blood Pressure," *Yoga Mimansa,* vol. 2, no. 2 (April 1926), second impression 1932, pp. 92–128.

28 *a pioneering set of measurements:* Anonymous (but clearly J. G. Gune), "Determination of CO_2 and O_2," *Yoga Mimansa,* vol. 4, no. 2 (November 1930), pp. 123–57; "O_2 Absorption and CO_2 Elimination in Pranayama," *Yoga Mimansa,* vol. 4, no. 4 (October 1933), pp. 267–89.

28 *"The idea that":* Anonymous (but clearly J. G. Gune), "Physiological and Spiritual Values of Pranayama," *Yoga Mimansa,* vol. 4, no. 4 (October 1933), p. 312.

28 *sent free copies:* Interview, O. P. Tiwari, secretary of Kaivalyadhama Yoga Ashram, Lonavla, India, June 27, 2007.

29 *a hero of the nationalist intelligentsia:* One tribute came in March 1930 from Motilal Nehru, founder of what would become India's most powerful political dynasty. The family produced three prime ministers. In a letter, Nehru said Gune had shown that yoga could withstand "the fierce light of modern sciences" and—in a dig at colonialism—found it to be "well in advance of all that has so far been discovered in the West." He said every Indian had a duty to support Gune "and afford him a full and fair opportunity to realize his ideals for the physical and cultural uplift of India." See Anonymous (but clearly J. G. Gune), "Editorial Notes: Pandit Motilal Nehru's Note," *Yoga Mimansa,* vol. 4, no. 1 (July 1930), p. 3.

29 *recommended the calming effect:* Gharote and Gharote, *Kuvalayananda,* pp. 57–58, 64–65, 96–99; Joseph S. Alter, *Gandhi's Body: Sex, Diet, and the Politics of Nationalism* (Philadelphia: University of Pennsylvania Press, 2000), pp. 18–19.

29 *yoga in classes of mass instruction:* Gharote and Gharote, *Kuvalayananda,* pp. 21, 88–95.

29 *"peculiarly fitted":* Anonymous (but clearly J. G. Gune), "Kaivalyadhâma— An Appeal," *Yoga Mimansa,* vol. 2, no. 4 (October 1926), second impression, December 1932, p. 294.

29 *"probably had a more profound impact":* Alter, *Gandhi's Body,* p. 68.

29 *played a skillful role:* James Manor, *Political Change in an Indian State: Mysore 1917–1955* (Columbia, MO: South Asia Books, 1978), pp. 1–27, 73–94; De Michelis, *A History,* pp. 196–97.

30 *practiced an eclectic style:* Norman E. Sjoman, *The Yoga Tradition of the Mysore Palace* (New Delhi: Abhinav Publications, 1999), pp. 55–62. The palace referred to advertising yoga as "propaganda work," see p. 50.

30 *hired a teacher:* Ibid., pp. 50–52, 109–10; De Michelis, *A History,* pp. 195–96.

30 *spent his early life:* Fernando Pagés Ruiz, "Krishnamacharya's Legacy," *Yoga Journal*, May–June 2001, pp. 96–101, 161–68.

30 *a style that drew:* Sjoman, *The Yoga Tradition*, p. 55.

30 *a number of gifted students:* They include T. K. V. Desikachar, Indra Devi, B. K. S. Iyengar, Pattabhi Jois, and Srivatsa Ramaswami. See Srivatsa Ramaswami, *The Complete Book of Vinyasa Yoga* (New York: Marlowe & Company, 2005), pp. xiii–xvi; Ruiz, "Krishnamacharya's Legacy."

30 *the maharajah asked Krishnamacharya:* B. K. S. Iyengar, *Iyengar: His Life and Work* (Porthill, ID: Timeless Books, 1987), p. 8.

30 Yoga Makaranda: T. N. Krishnamacariar, *Yoga Makaranda* (Mysore: Mysore Palace, 1935).

30 *"pains in the abdomen":* Quoted in Sjoman, *The Yoga Tradition*, p. 66.

31 *sickly all his life:* B. K. S. Iyengar, *Astadala Yogamala* (New Delhi: Allied Publishers, 2006), vol. 1, pp. 15–16.

31 *afterward helped facilitate:* Ibid., pp. 27–28; see also De Michelis, *A History*, pp. 197–98.

31 *a knowledgeable liaison:* See Anne Cushman, "Iyengar Looks Back," *Yoga Journal*, November–December 1997, pp. 85–91, 156–65.

32 *misalignments that could restrict:* For a description, see Robin, *A Physiological Handbook*, p. 249.

32 *turn the foot ninety degrees:* Iyengar, *Light on Yoga*, pp. 63–64.

33 *saw Iyengar perform:* Iyengar, *Iyengar*, p. 25.

33 *studied at his ashram:* Indra Devi, *Forever Young, Forever Healthy: Simplified Yoga for Modern Living* (New York: Prentice Hall, 1953), pp. 15–17.

33 *"He said he had no classes":* Ibid., p. 18.

34 *experience with yogic supermen:* Paramahansa Yogananda, *Autobiography of a Yogi* (Los Angeles: Self-Realization Fellowship, 1994).

34 *"Control over death":* Paramahansa Yogananda, *Scientific Healing Affirmations: Theory and Practice of Concentration* (Los Angeles: Self-Realization Fellowship, 1998), p. 29.

34 *"Yogis know how":* Swami Yogananda, *Super Advanced Course Number 1, Lessons 1 to 12 (1930)* (Whitefish, MT: Kessinger, 2003), p. 8.

34 *a close friendship with Yogananda:* Swami Satyananda Giri, *A Collection of Biographies of 4 Kriya Yoga Gurus* (New York: iUniverse, 2004), pp. 184, 187–202, 220–32, 266–67; Laurel Elizabeth Keyes, *Sundial* (Denver: Gentle Living Publications, 1981), pp. 99–105.

34 *the second-in-command:* "I am powerless," Yogananda wrote in 1925, "to tell how greatly he has helped me." Swami Yogananda, "Swami Dhirananda," *East West,* vol. 1, no. 1 (November–December 1925), p. 29. Dhirananda was Bagchi's monastic name.

34 *breaking his vow of celibacy:* Ron Russell, "The Devotee's Son," *New Times Los Angeles,* July 5, 2001.

35 *a pioneer:* Anonymous, "B. K. Bagchi: Memoir," September Meeting 1965, *Proceedings of the Board of Regents: July 26, 1963–June 23, 1965* (Ann Arbor: University of Michigan Press, n.d.), pp. 1016–17.

35 *his death reported:* Anonymous, "Speaker Dies While Introducing Indian Ambassador at Dinner Here," *Los Angeles Times,* March 8, 1952, p. 1.

35 *had taken to demonstrating:* Ruiz, "Krishnamacharya's Legacy"; Iyengar, *Iyengar,* pp. 15, 17.

35 *the beat was still there:* Today, doctors call what Krishnamacharya did the Valsalva maneuver. When a person breathes deeply and holds their breath, the strain creates pressure in the chest that traps blood in the venous system and slows its flow into the heart or blocks it altogether, diminishing the heartbeat. See William D. McArdle, Frank I. Katch, and Victor L. Katch, *Exercise Physiology: Nutrition, Energy, and Human Performance,* 6th ed. (Philadelphia: Lippincott Williams & Wilkins, 2007), pp. 272–75.

35 *published their findings:* M. A. Wenger, B. K. Bagchi, and B. K. Anand, "Experiments in India on 'Voluntary' Control of the Heart and Pulse," *Circulation,* vol. 24 (December 1961), pp. 1319–25.

35 *"It was often reported":* Basu K. Bagchi, "Mysticism and Mist in India," *Journal of the American Society of Psychosomatic Dentistry and Medicine,* vol. 16, no. 3 (1969), pp. 73–87.

36 *more than five weeks:* Gharote and Gharote, *Kuvalayananda,* p. 77.

36 *measured six feet long:* M. V. Bhole, P. V. Karambelkar, and S. L. Vinekar, "Underground Burial or Bhugarbha Samadhi (Part II)," *Yoga Mimansa,* vol. 10, no. 2 (October 1967), pp. 1–16.

36 *an itinerant showman:* Arthur Koestler, *The Lotus and the Robot* (London: Hutchinson, 1960), pp. 116–19. Koestler here recounts an early phase of the

experiment before the pit walls were tightly sealed, though the showman participated in all stages of the study.

36 *Twice in 1962:* Bhole et al., "Underground Burial." In the table, the subject, Ramandana Yogi, is referred to by the initials RN.

37 *locked volunteers into the samadhi pit:* Yoga turned out to confer no advantage. The experiments showed that a person in the pit tended to breathe less oxygen. But the cause, the scientists found, was simply the body's natural response to rising levels of carbon dioxide as exhaled air accumulated. The high levels lowered the body's overall metabolism and thus the need for oxygen. Their observations jibed with what Paul had described a century earlier. The cooling of the body's fires, they wrote, "is neither voluntary in nature" nor "controlled by Yogic methods." A fervent skeptic could have hardly said it more firmly. See Bhole et al., "Underground Burial."

37 *"We're still ready":* Interview, Makrand Gore, Kaivalyadhama Yoga Ashram, Lonavla, India, June 28, 2007.

37 *"I can drink acid":* Iyengar, *Light on Yoga*, p. 13.

37 *"relieves pain":* Ibid., p. 100.

38 *"makes healthy pure blood":* Ibid., p. 190.

38 *the possibility of placebo effects:* For how they combine with poor experimental design to produce false conclusions, see R. Barker Bausell, *Snake Oil Science: The Truth About Complementary and Alternative Medicine* (New York: Oxford University Press, 2007).

38 *At the book's end:* Iyengar, *Light on Yoga*, pp. 487–506.

39 *"sexual retentive power":* Ibid., p. 438.

40 *investigators at the University of Pennsylvania:* Debbie L. Cohen, LeAnne T. Bloedon, Rand L. Rothman, et al., "Iyengar Yoga versus Enhanced Usual Care on Blood Pressure in Patients with Prehypertension to Stage I Hypertension: A Randomized Controlled Trial," eCAM, Oxford University Press, September 4, 2009, pp. 1–8.

40 *dozens conducted everywhere:* Kim E. Innes, Cheryl Bourguignon, and Ann Gill Taylor, "Risk Indices Associated with the Insulin Resistance Syndrome, Cardiovascular Disease, and Possible Protection with Yoga: A Systematic Review," *Journal of the American Board of Family Medicine*, vol. 18, no. 6 (November–December 2005), pp. 491–519.

40 *"safe and cost-effective":* Ibid., p. 492.

41 *Consider a 2011 study:* Kathleen K. Zettergren, Jennifer M. Lubeski, and Jaclyn M. Viverito, "Effects of a Yoga Program on Postural Control, Mobility, and Gait Speed in Community-Living Older Adults: A Pilot Study," *Journal of Geriatric Physical Therapy*, vol. 34, no. 2 (April–June 2011), pp. 88–94.

41 *counteract the deterioration of the disks:* Robin, *A Handbook*, pp. 279–80.

41 *With normal aging:* Michel Benoist, "Natural History of the Aging Spine," *European Spine Journal*, vol. 12, supplement 2 (2003), pp. S86–S89.

41 *a study of thirty-six people:* Chin-Ming Jeng, Tzu-Chieh Cheng, Ching-Huei Kung, et al., "Yoga and Disc Degenerative Disease in Cervical and Lumbar Spine: An MR Imaging-based Case Control Study," *European Spine Journal*, vol. 20, no. 3 (March 2011), pp. 408–13.

42 *reported that the vagus:* Kevin J. Tracey, "The Inflammatory Reflex," *Nature*, vol. 420, no. 6917 (December 19–26, 2002), pp. 853–59.

42 *two hundred thousand lives:* Roni Caryn Rabin, "Awareness: Killer of 200,000 Americans, Hardly Noticed," *New York Times*, October 4, 2010, Section D, p. 6.

42 *discussed the topic:* Stacey L. Oke and Kevin J. Tracey, "The Inflammatory Reflex and the Role of Complementary and Alternative Medical Therapies," in William C. Bushell, Erin L. Olivo, and Neil D. Theise, eds., *Longevity, Regeneration, and Optimal Health: Integrating Eastern and Western Perspectives* (New York: Blackwell for the New York Academy of Sciences, 2009), pp. 172–80.

42 *ease trauma from rheumatoid arthritis:* Shirley Telles, Kalkuni V. Naveen, Vaishali Gaur, et al., "Effect of One Week of Yoga on Function and Severity in Rheumatoid Arthritis," *BMC Research Notes*, vol. 4 (2011), p. 118.

43 *reported that yogis could live:* Eliade, *Yoga*, p. 275.

43 *no study that I know of:* The closest thing I could find is a paper on the effects of Transcendental Meditation. See Robert H. Schneider, Charles N. Alexander, Frank Staggers, et al., "Long-Term Effects of Stress Reduction on Mortality in Persons > 55 Years of Age with Systemic Hypertension," *American Journal of Cardiology*, vol. 95, no. 9 (May 1, 2005), pp. 1060–64.

43 *won the 2009 Nobel Prize:* Nicholas Wade, "3 Americans Share Nobel for Medicine," *New York Times*, October 5, 2009, Section A, p. 12.

43 *youthful telomeres of a thirty-year-old:* Elizabeth H. Blackburn, "Telomeres and Telomerase: The Means to the End," Nobel Lecture, December 7, 2009. Blackburn was one of three winners of the 2009 Nobel Prize.

43 *slow the biological clock:* Thea Singer, *Stress Less: The New Science That Shows Women How to Rejuvenate the Body and the Mind* (New York: Hudson Street Press, 2010), pp. xviii–xix, 29–62. For a skeptical look at the science of Telomere evaluation; see Mitch Leslie, "Are Telomere Tests Ready for Prime Time?" *Science,* April 22, 2011, pp. 414–15.

44 *a longtime devotee of yoga:* Judith Lasater, "Yoga and Your Heart," *Yoga Journal,* September–October 1989, pp. 13–15.

44 *proclaiming them a first:* Dean Ornish, Jue Lin, Jennifer Daubenmier, et al., "Increased Telomerase Activity and Comprehensive Lifestyle Changes: A Pilot Study," *Lancet Oncology,* vol. 9, no. 11 (November 2008), pp. 1048–57.

45 *"live to be well over 100 years":* Joan Budilovsky and Eve Adamson, *The Complete Idiot's Guide to Yoga,* 3rd ed. (Indianapolis: Alpha Books, 2003), p. 10.

45 *"lent credibility":* Georg Feuerstein, *Sacred Paths: Essays on Wisdom, Love, and Mystical Realization* (Burdett, NY: Larson Publications, 1991), p. 50.

II: Fit Perfection

47 *hundreds of yoga centers:* Email, Ainslie Faust, director of communications, Bikram's Yoga College of India, International Headquarters, Los Angeles, July 17, 2010. "About 500 and many more illegal," she said of the number of studios globally. This is less than a third of the 1,700 studios that Bikram Choudhury claims on the jacket of his book *Bikram Yoga: The Guru Behind Hot Yoga Shows the Way to Radiant Health and Personal Fulfillment* (New York: HarperCollins, 2007). What accounts for the discrepancy is unclear.

47 *calls it his torture chamber:* Ibid., pp. 67, 73–74, 76, 96, 215.

47 *"So many Americans":* Ibid., p. 45.

48 *"Bogus yoga":* Ibid., pp. 61–67.

48 *portrays his own style:* Ibid., p. 5.

50 *"My classes are so hard":* Quoted in Nancy Keates, "Is Yoga Just Posing as a Good Workout?" *Wall Street Journal,* November 17, 2007, Section W, p. 1.

50 *was seen as urgent:* Gina Kolata, *Ultimate Fitness: The Quest for Truth About Exercise and Health* (New York: Picador, 2004), pp. 36–39.

50 *vital capacity:* Guy Montrose Whipple, *Manual of Mental and Physical Tests* (Baltimore: Warwick & York, 1910), pp. 70–74.

51 *"increasing the vital index":* Anonymous (but clearly J. G. Gune), "The Rationale of Yogic Poses," *Yoga Mimansa*, vol. 3, no. 2 (April 1928), second impression, January 1931, pp. 121–26.

51 *an English physiologist:* Anonymous, "Archibald V. Hill: Biography," The Nobel Foundation, nobelprize.org/nobel_prizes/medicine/laureates/1922 /hill-bio.html.

52 *an abiding personal interest:* David R. Bassett, Jr., "Scientific Contributions of A. V. Hill: Exercise Physiology Pioneer," *Journal of Applied Physiology*, vol. 93 (November 2002), pp. 1567–82.

52 *In pioneering reports:* For a review, see David R. Bassett, Jr., and Edward T. Howley, "Limiting Factors for Maximum Oxygen Uptake and Determinants of Endurance Performance," *Medicine & Science in Sports & Exercise*, vol. 32, no. 1 (January 2000), pp. 70–84.

52 *the single most important factor:* Benjamin D. Levine, "VO_2 Max: What Do We Know, and What Do We Still Need to Know," *Journal of Physiology*, vol. 586, no. 1 (January 1, 2008), pp. 25–34.

53 *twice that of untrained individuals:* Joe Warpeha, "Limitation of Maximal Oxygen Consumption: The Holy Grail of Exercise Physiology or Fool's Gold?" *Professionalization of Exercise Physiology—online*, vol. 6, no. 9 (September 2003).

53 *could raise VO_2 max:* Jay Hoffman, *Physiological Aspects of Sport Training and Performance* (Champaign, IL: Human Kinetics, 2002), p. 111.

53 *Cooper came along:* Kolata, *Ultimate Fitness*, pp. 25–29, 45–46.

53 *the best cardiovascular workout:* Kenneth H. Cooper, *Aerobics* (New York: Bantam, 1968), pp. 15–26.

54 *reduced the prevalence:* Jane E. Brody, "You Name It, and Exercise Helps It," *New York Times*, April 29, 2008, Section F, p. 7.

54 *"the single thing":* Quoted in Jonathan Shaw, "The Deadliest Sin," *Harvard Magazine*, March–April 2004, pp. 36–43, 98–99.

54 *at least three vigorous exercise sessions:* Lisa K. Lloyd, "Are You Ready to Exercise?" *Fit Society Page*, American College of Sports Medicine, Summer 2001, p. 1.

54 *at least five sessions:* William L. Haskell, I-Min Lee, Russell R. Pate, et al., "Physical Activity and Public Health: Updated Recommendation for Adults from the American College of Sports Medicine and the American Heart Association," *Circulation*, vol. 116, no. 9 (August 28, 2007), pp. 1081–93.

54 *the President's Council:* Anonymous, "Fitness Fundamentals: Guidelines for Personal Exercise Programs," The President's Council on Physical Fitness, undated, www.fitness.gov/fitness.htm.

54 *"reducing cardiovascular diseases":* Anonymous, "Diet, Nutrition and the Prevention of Chronic Diseases," World Health Organization, WHO Technical Report No. 916, 2003, pp. 62–63.

55 *quibble over the amounts:* Kolata, *Ultimate Fitness,* pp. 51–72.

55 *carefully examined several activities:* Cooper, *Aerobics,* pp. 15–26.

56 *published in 1989:* James A. Blumenthal, Charles F. Emery, David J. Madden, et al., "Cardiovascular and Behavioral Effects of Aerobic Exercise Training in Healthy Older Men and Women," *Journal of Gerontology,* vol. 44, no. 5 (September 1989), pp. M147–M157.

58 *makes no appearance:* Kuvalayananda, *Popular Yoga;* Sri Swami Sivananda, *Easy Steps to Yoga,* reprint of 1939 edition (Tehri-Garhwal, Uttar Pradesh, India: Divine Life Society, 1999).

58 *The pose most likely arose:* Sjoman, *The Yoga Tradition,* pp. 50, 54, 58; Anne Cushman, "New Light on Yoga," *Yoga Journal,* July–August 1999, pp. 44–49; Joseph S. Alter, *The Wrestler's Body: Identity and Ideology in North India* (Berkeley: University of California Press, 1992), pp. 98–105; Vincent Giordano, *The Physical Body: Indian Wrestling and Physical Culture,* video disc, CustomFlix, 2006.

58 *spreading slowly through India:* For a case study of its adoption by the Raja of Aundh, the ruler of a diminutive state north of Mysore, see Katherine H. Diver, *Royal India: A Descriptive and Historical Study of India's Fifteen Principal States and Their Rulers* (London: Hodder & Stoughton, 1942), pp. 148–50.

58 *learned about the pose:* In 1959, a popular book by Swami Vishnudevananda, a young disciple of Swami Sivananda's, gained wide notice. It was filled with dozens of photographs of the young swami and clear, how-to directions, including eleven pages devoted to the Sun Salutation. See Swami Vishnudevananda, *The Complete Illustrated Book of Yoga* (New York: Bell Publishing, 1960).

59 *examined sixteen volunteers:* Virginia S. Cowen and Troy B. Adams, "Heart Rate in Yoga Asana Practice: A Comparison of Styles," *Journal of Bodywork and Movement Therapies,* vol. 11, no. 1 (January 2007), pp. 91–95.

59 *"the lack of objective study":* Email to author, Ezra A. Amsterdam, October 22, 2011.

60 *"many lively discussions":* Email to author, Dina Amsterdam, October 26, 2011. For background on Amsterdam's yoga study and Dina's role, see Alisa Bauman, "Is Yoga Enough to Keep You Fit?" *Yoga Journal,* September–October 2002, p. 84.

60 *just ten volunteers:* Mark D. Tran, Robert G. Holly, Jake Lashbrook, and Ezra A. Amsterdam, "Effects of Hatha Yoga Practice on the Health-Related Aspects of Physical Fitness," *Preventive Cardiology,* vol. 4, no. 4, (Fall 2001), pp. 165–70.

61 *Different schools of yoga:* The Iyengar Frog is a static posture. See Iyengar, *Light on Yoga,* pp. 126–28. The energetic Frog of the Davis study is taught by the Kundalini school. See Shakta Kaur Khalsa, *Kundalini Yoga: Unlock Your Inner Potential Through Life-Changing Exercise* (New York: Dorling Kindersley, 2001), pp. 55, 72, 100, 129.

62 *making their evaluation "blind":* Emails to author, Ezra A. Amsterdam, October 22 and 23, 2011. Amsterdam said reviewers for the journal once rejected a paper on which he was listed as a coauthor, showing how the process was unbiased.

64 *"the most current scientific information":* Anonymous, "The Yoga Journal Story," www.yogajournal.com/global/34.

64 *"Yoga is all you need":* Bauman, "Is Yoga Enough to Keep You Fit?"

64 *a commercial style:* Beth Shaw, *Beth Shaw's YogaFit* (Champaign, IL: Human Kinetics, 2000), pp. ix–xvi; Beth Greenfield, "Strike a Pose," *Out,* February 2004, p. 78.

65 *developed a course of training:* Kolata, *Ultimate Fitness,* p. 247.

65 *"a tough cardiovascular workout":* Shaw, *Beth Shaw's,* p. ix.

65 *Her promotional literature:* Anonymous, "Host a Yoga Instructor Training," www.yogafit.com/host-a-training.shtml.

65 *The sixteen-page paper:* Chrys Kub, "Health Benefits of Hatha Yoga," www.yogafit.com/research/healthbenefits.doc.

65 *"You don't need":* Alexa Joy Sherman, "Total Body Power Yoga," *Shape,* vol. 24, no. 3 (November 2004), pp. 186–91.

66 *a young scientist who practiced yoga:* Emails to author, Carolyn C. Clay, Texas State University, June 9 and 15, 2011.

66 *Their study appeared:* Carolyn C. Clay, Lisa K. Lloyd, John L. Walker, et al., "The Metabolic Cost of Hatha Yoga," *The Journal of Strength and Conditioning Research,* vol. 19, no. 3 (August 2005), pp. 604–10.

68 *more emphasis on the vigorous pose:* For a scientific evaluation of the Sun Salutation as an exclusive path to cardiovascular fitness, see Bhavesh Surendra Mody, "Acute effects of Surya Namaskar on the cardiovascular & metabolic system," *Journal of Bodywork and Movement Therapies*, vol. 15, no. 3 (July 2011), pp. 343–47.

68 *done at the University of Wisconsin:* Dawn D. Boehde, John P. Porcari, John Greany, et al., "The Physiological Effects of 8 Weeks of Yoga Training," *Journal of Cardiopulmonary Rehabilitation*, vol. 25 (2005), p. 290.

68 *"The intensity just wasn't there":* Quoted in Mark Anders, "Does Yoga Really Do the Body Good?" American Council on Exercise, *ACE Fitness Matters*, September–October 2005, pp. 7–9.

68 *sent out a press release:* Anonymous, "ACE First to Evaluate Benefits of Yoga," the American Council on Exercise, September 28, 2005, www.acefit ness.org/pressroom/419/ace-first-to-evaluate-benefits-of-yoga-br-i.

69 *"Aerobics?":* John Briley, "Aerobics? Not Among Yoga's Strengths," *Washington Post*, October 11, 2005, p. F3.

69 Yoga Journal *took notice:* Sierra Senyak, "Flexible *and* Fit," *Yoga Journal*, June 2006, p. 22.

69 *"I think you just proved":* Mia, reader comment on "Flexible and Fit," www .yogajournal.com/lifestyle/2275.

69 *A final study:* Marshall Hagins, Wendy Moore, and Andrew Rundle, "Does Practicing Hatha Yoga Satisfy Recommendations for Intensity of Physical Activity Which Improves and Maintains Health and Cardiovascular Fitness?" *BMC Complementary and Alternative Medicine*, vol. 7, no. 40 (November 30, 2007), www.biomedcentral.com/1472-6882/7/40.

70 *The lead scientist:* Anonymous, "Despite Other Health Benefits, Yoga May Not Lead to Cardio Fitness Says Researcher at Long Island University's Brooklyn Campus in New Study," news release, Brooklyn campus of Long Island University, January 8, 2008, www.brooklyn.liu.edu/wn/2008/004 .html.

70 *Known as a metabolic chamber:* Interview, Marshall Hagins, Long Island University, July 21, 2010.

73 *examined more than eighty studies:* Alyson Ross and Sue Thomas, "The Health Benefits of Yoga and Exercise: A Review of Comparison Studies," *Journal of Alternative and Complementary Medicine*, vol. 16, no. 1 (January 2010), pp. 3–12.

74 *"a good cardio workout"*: Anonymous, "What's Your Style, Baby?" *Yoga Journal*, February 2008, p. A7.

74 *"aerobic benefits"*: Georg Feuerstein and Larry Payne, *Yoga for Dummies*, 2nd ed. (Indianapolis: Wiley, 2010), p. 172.

74 *a frequent question of readers*: Ann Pizer, "Does Yoga Keep You Fit?" About.com, December 11, 2008, http://yoga.about.com/b/2008/12/11/does-yoga-keep-you-fit.htm.

74 *advised readers that vigorous styles*: Michael J. O'Fallon and Denney G. Rutherford, *Hotel Management and Operations*, 5th ed. (Hoboken, NJ: Wiley, 2010), p. 41.

74 *Huffington Post ran a link*: Verena von Pfetten, "Is Yoga Really Enough To Keep You Fit?," Huffington Post, July 31, 2008, www.huffingtonpost.com/2008/07/31/is-yoga-really-enough-to_n_116158.html.

75 *asked its readers in 2010*: Ivy Markaity, "Does Yoga Provide Enough of a Cardio Workout?" HealthCentral.com, August 27, 2010, www.healthcentral.com/diet-exercise/c/223360/118686/workout.

75 *runs a school in Los Angeles*: Lee Jenkins, "Deep Breath as Pitchers Rethink Routines," *New York Times*, February 12, 2007, Section D, p. 1.

III: Moods

78 *"What ever happened"*: Interview, Sat Bir Khalsa, Harvard Medical School, May 15, 2007.

79 *more important than money*: Robert E. Thayer, *The Origin of Everyday Moods* (New York: Oxford University Press, 1996), pp. 3–42.

79 *first definitions*: The Compact Edition of The Oxford English Dictionary (Oxford: Oxford University Press, 1971), vol. 1, p. 1844; Albert C. Baugh and Thomas Cable, *A History of the English Language* (London: Routledge, 1993), p. 63.

79 *The conventional wisdom*: American Psychiatric Association, the *Diagnostic and Statistical Manual of Mental Disorders*, 4th ed. (Washington: American Psychiatric Association, 1994), p. 341.

79 *A survey found*: Laura A. Pratt, Debra J. Brody, and Qiuping Gu, "Antidepressant Use in Persons Aged 12 and Over: United States, 2005–2008," U.S. Department of Health and Human Services, Centers for Disease Control and Prevention, National Health and Nutrition Examination Surveys, Data Brief No. 76, October 2011.

79 *"It really saved"*: Yoga class with Amy Weintraub, Kripalu Yoga Center, Stockbridge, Massachusetts, November 12, 2010.

80 *"as through a fog"*: Amy Weintraub, *Yoga for Depression: A Compassionate Guide to Relieve Suffering Through Yoga* (New York: Broadway Books, 2004), p. 2.

80 *a book in the works*: Amy Weintraub, *Yoga Skills for Therapists: Mood-Management Techniques to Teach & Practice* (New York: Norton, 2012).

80 *core attraction of Kripalu*: For an overview of the center, see Andy Newman, "It's Not Easy Picking a Path to Enlightenment," *New York Times*, July 3, 2008, Section G, p. 1.

80 *"Free in this world"*: Quoted in David H. Albert, "Thoreau's India: The Impact of Reading in a Crisis," *Proceedings of the American Philosophical Society*, vol. 125, no. 2 (April 30, 1981), p. 107.

80 *the first known instance*: De Michelis, *A History*, pp. 2–3.

81 *looked favorably on yoga*: William James, *On Vital Reserves: The Energies of Men, The Gospel of Relaxation* (New York: Henry Holt, 1922), pp. 26–29.

81 *simple but systematic relaxation*: William James, *The Varieties of Religious Experience* (New York: Longmans, Green, 1902), pp. 110–15.

81 *following the lead of James*: F. J. McGuigan and Paul M. Lehrer, "Progressive Relaxation: Origins, Principles, and Clinical Application," in Paul M. Lehrer, Robert L. Woolfolk, Wesley E. Sime, et al., eds., *Principles and Practice of Stress Management*, 3rd ed. (New York: Guilford Press, 2007), pp. 57–87.

81 *subjects had no obvious reaction*: Edmund Jacobson, *Progressive Relaxation: A Physiological and Clinical Investigation of Muscular States and Their Significance in Psychology and Medical Practice* (Chicago: University of Chicago Press, 1931), pp. 101–11.

82 *taught himself how to relax*: Ibid., p. 111.

82 *produced remarkable cures*: Ibid., pp. 309–79.

82 *suffered a skull fracture*: Edmund Jacobson, "The Neurovoltmeter," *American Journal of Psychology*, vol. 52, no. 4 (October 1939), pp. 620–24.

83 *graduated from the University*: Kovoor T. Behanan, *Yoga: A Scientific Evaluation* (New York: Macmillan, 1937), pp. xii–xiii.

83 *journeyed to the world capital*: The arrangement carried a risk of conflict, given that Gune was a true believer and Behanan an outsider committed,

in theory, to scientific doubt. But Gune was gracious, and Yale politic. Behanan's advisor, Walter R. Miles, a professor at Yale's Institute of Human Relations, said Gune saw his new student as having a "sympathetic and genuine desire to study the system of Yoga objectively and critically." See ibid, p. xiv.

83 *practiced every day:* Ibid., p. 229.

83 *did the exercise at a rate:* Ibid., p. 237.

83 *psychological testing:* Ibid., pp. 228–32.

84 Life *ran a formal portrait:* Anonymous, "Speaking of Pictures . . . These are the Exercises of Yoga," *Life*, April 19, 1937, pp. 8–9.

84 *a glowing review:* Anonymous, "Yale's Yogin," *Time*, April 26, 1937, p. 24

84 *"a retardation of mental functions":* Behanan, *Yoga*, p. 232.

84 *"an extremely pleasant feeling":* Ibid., p. 243.

84 *"emotional stability":* Ibid., p. 245.

84 *a spike in oxygen consumption:* Ibid., p. 237.

85 *exploits this ocean:* McArdle et al., *Exercise Physiology*, pp. 282–85.

85 *nearly saturated with oxygen:* Richard M. Schwartzstein and Michael J. Parker, *Respiratory Physiology: A Clinical Approach* (Philadelphia: Lippincott Williams & Wilkins, 2006), p. 109.

85 *large quantities no matter what:* An exception arises when the lungs are diseased or impaired, at which point slow breathing may aid oxygenation. See Luciano Bernardi, Giammario Spadacini, Jerzy Bellwon. et al., "Effect of Breathing Rate on Oxygen Saturation and Exercise Performance in Chronic Heart Failure," *Lancet*, vol. 351 (May 2, 1998), pp. 1308–11.

85 *creating an inner environment:* McArdle et al., *Exercise Physiology*, pp. 280–81.

86 *"lowers body stores":* U.S. Navy, *U.S. Navy Diving Manual*, Revision Five, vol. 1, (Washington, DC: Naval Sea Systems Command, 2005), p. 3–19.

86 *respiratory alkalosis:* Schwartzstein and Parker, *Respiratory Physiology*, pp. 155–56; Carol Mattson Porth, *Essentials of Pathophysiology* (Philadelphia: Lippincott Williams & Wilkins, 2007), p. 147; Burton David Rose and Theodore W. Post, *Clinical Physiology of Acid-Base and Electrolyte Disorders* (New York: McGraw-Hill, 2001), pp. 673–81.

86 *"a feeling of exhilaration":* Iyengar, *Light on Pranayama*, p. 179.

86 *vessels in the brain to contract:* Ronald Ley, "Breathing and the Psychology of Emotion, Cognition, and Behavior," in Beverly H. Timmons and Ronald Ley, eds., *Behavioral and Psychological Approaches to Breathing Disorders* (New York: Kluwer, 1994), pp. 86–88; L. C. Lum, "Hyperventilation Syndromes," in Timmons and Ley, *Behavioral*, pp. 119–20.

86 *cuts levels roughly in half:* Ibid., p. 119; for an early study, see Seymour S. Kety and Carl F. Schmidt, "The Effects of Active and Passive Hyperventilation on Cerebral Blood Flow, Cerebral Oxygen Consumption, Cardiac Output, and Blood Pressure of Normal Young Men," *Journal of Clinical Investigation*, vol. 25 (1946), pp. 107–19.

87 *three scientists in Sweden:* C. Frostell, J. N. Pande, and G. Hedenstierna, "Effects of high-frequency breathing on pulmonary ventilation and gas exchange," *Journal of Applied Physiology*, vol. 55, no. 6 (1983), pp. 1854–61.

88 *a standard figure:* Porth, *Essentials*, p. 515.

88 *brain now gets more oxygen:* Ibid., p. 147.

88 *increases in calm alertness:* Richard P. Brown and Patricia L. Gerbarg, "Sudarshan Kriya Yogic Breathing in the Treatment of Stress, Anxiety, and Depression: Part I—Neurophysiologic Model," *Journal of Alternative and Complementary Medicine*, vol. 11, no. 1 (2005), pp. 189–201.

88 *linked slow breathing to heightened vigilance:* Dirk S. Fokkema, "The Psychobiology of Strained Breathing and Its Cardiovascular Implications: A Functional System Review," *Psychophysiology*, vol. 36, no. 2 (March 1999), pp. 164–75.

88 *a study of nearly two dozen adults:* Luciano Bernardi, Peter Sleight, Gabriele Bandinelli, et al., "Effect of Rosary Prayer and Yoga Mantras on Autonomic Cardiovascular Rhythms: A Comparative Study," *British Medical Journal*, vol. 323 (2001), pp. 1446–49.

88 *"You're not used":* Choudhury, *Bikram Yoga*, p. 101.

88 *Kundalini Yoga:* Khalsa, *Kundalini Yoga*, p. 25.

88 *"one of the best things":* Budilovsky and Adamson, *The Complete Idiot's Guide*, p. 302.

89 *"for anyone in need":* Anonymous, "Yoga Instructional Book 'Oxygen Yoga: A Spa Universe' Now Available," WiredPRNews.com, August 15, 2011.

89 *wrote extensively:* Her writings were often in French. For an English work, see Thérèse Brosse, "A Psycho-physiological Study," *Main Currents in Modern Thought*, no. 4 (July 1946), pp. 77–84.

89 *"an extreme slowing"*: B. K. Bagchi and M. A. Wenger, "Electro-physiological Correlates of Some Yogi Exercises," *EEG and Clinical Neurophysiology*, 1957, supplement 7, p. 146.

90 *a gap of up to eleven degrees:* Elmer and Alyce Green, *Beyond Biofeedback* (New York: Delacorte Press, 1977), pp. 197–205.

90 *biologists called it "sympathetic":* Stanley Finger, *Origins of Neuroscience: A History of Explorations into Brain Function* (New York: Oxford University Press, 2001), pp. 280–84.

91 *"My name is Mel":* Yoga class with Mel Robin, Yoga Loft, Bethlehem, Pennsylvania, April 14, 2007.

93 *cast the topic:* Yoga class with Mel Robin, Yoga Loft, Bethlehem, Pennsylvania, July 24, 2007.

94 *a place in his book:* Robin, *A Physiological Handbook*, pp. 137–40.

94 *had us turn to another page:* Ibid., p. 147.

95 *The Relaxation Response*: Herbert Benson and Miriam Z. Klipper, *The Relaxation Response* (New York: HarperCollins, 2000).

96 *average of 13 percent:* Chaya et al., "The Effect of Long Term Combined Yoga Practice." See also M. S. Chaya and H. R. Nagendra, "Long-term effect of yogic practices on diurnal metabolic rates of healthy subjects," *International Journal of Yoga*, vol. 1, no. 1 (January–June 2008), pp. 27–32.

97 New York Times *profiled:* Lizette Alvarez, "Rebel Yoga," *New York Times*, January 23, 2011, Section WE, p. 1.

97 *"rev up your metabolism":* Tara Stiles, *Slim Calm Sexy Yoga* (New York: Rodale, 2010), p. 150.

98 *"Yoga affects the mind":* Interview, Mayasandra S. Chaya, Swami Vivekananda Yoga Research Foundation, Bangalore, India, June 13, 2011.

98 *nearly one million despairing people:* Anonymous, "Depression," World Health Organization.

99 *known about GABA since the 1950s:* Max R. Bennett, *History of the Synapse* (Amsterdam: Harwood Academic Publishers, 2001), pp. 83–85.

100 *published in 2007:* Chris C. Streeter, J. Eric Jensen, Ruth M. Perlmutter, et al., "Yoga Asana Sessions Increase Brain GABA Levels: A Pilot Study," *Journal of Alternative and Complementary Medicine*, vol. 13, no. 4 (2007), pp. 419–26.

100 *"clear public health advantages":* Quoted in Gina M. Digravio, "Study Finds Yoga Associated with Elevated Brain GABA Levels," Boston University news release, May 22, 2007.

100 *published in 2010:* Chris C. Streeter, Theodore H. Whitfield, Liz Owen, et al., "Effects of Yoga Versus Walking on Mood, Anxiety, and Brain GABA Levels: A Randomized Controlled MRS Study," *Journal of Alternative and Complementary Medicine,* vol. 16, no. 11 (2010), pp. 1145–52.

101 *her website advised:* Liz Owen Yoga, www.lizowenyoga.com.

101 *reported in their 2006 paper:* Sat Bir S. Khalsa and Stephen Cope, "Effects of a Yoga Lifestyle Intervention on Performance-Related Characteristics of Musicians: A Preliminary Study," *Medical Science Monitor,* vol. 12, no. 8 (August 2006), pp. CR325–CR331.

102 *Khalsa and colleagues reported:* Sat Bir S. Khalsa, Stephanie M. Shorter, Stephen Cope, et al., "Yoga Ameliorates Performance Anxiety and Mood Disturbance in Young Professional Musicians," *Applied Psychophysiology and Biofeedback,* vol. 34, no. 4 (December 2009), pp. 279–89.

IV: Risk of Injury

103 *"Real yoga is as safe":* Swami Gitananda Giri, "Real Yoga Is as Safe as Mother's Milk," *Yoga Life,* vol. 28, no. 12 (December 1997), pp. 3–12.

104 *challenged the reports as biased:* See, for instance, Enoch Haga, "Yoga," *Journal of the American Medical Association,* vol. 218, no. 1 (October 4, 1971), p. 98.

104 *feature lengthy addendums:* Robin, *A Physiological Handbook,* pp. 511–18; Robin, *A Handbook,* pp. 833–41.

106 *"I make it as hard as possible":* Yoga class with Glenn Black, Sankalpah Yoga, New York City, January 24, 2009.

110 *reports began to emerge:* One of the earliest, if not the first, is Gilbert E. Corrigan, "Fatal Air Embolism after Yoga Breathing Exercises," *Journal of the American Medical Association,* vol. 210, no. 10 (December 8, 1969), p. 1923.

110 *an examination showed:* Joseph Chusid, "Yoga Foot Drop," *Journal of the American Medical Association,* vol. 217, no. 6 (August 9, 1971), pp. 827–28.

110 *in Vajrasana had clamped:* Robin, *A Physiological Handbook,* p. 513.

111 *chanting while standing:* Anonymous, "The Yoga Ailment," *Time*, August 23, 1971, p. 52.

111 *similar cases emerged:* See, for instance, Caryn M. Vogel, Roger Albin, and James W. Albers, "Lotus Footdrop: Sciatic Neuropathy in the Thigh," *Neurology*, vol. 41, no. 4 (April 1991), pp. 605–6; Thomas G. Mattio, Takashi Nishida, and Michael M. Minieka, "Lotus Neuropathy: Report of a Case," *Neurology*, vol. 42 (August 1992), p. 1636.

111 *One of the worst:* Melanie Walker, Gregg Meekins, and Shu-Ching Hu, "Yoga Neuropathy: A Snoozer," *Neurologist*, vol. 11, no. 3 (May 2005), pp. 176–78.

112 *published his pioneering research:* See, for example, D. Denny-Brown and W. Ritchie Russell, "Experimental Cerebral Concussion," *Brain*, vol. 64 (1941), pp. 93–164.

112 *His new warning:* W. Ritchie Russell, "Yoga and the Vertebral Arteries," *British Medical Journal*, vol. 1, no. 5801 (March 11, 1972), p. 685.

113 *typically move the vertebrae:* Robin, *A Physiological Handbook*, pp. 72–75. Extreme movements of the neck can, short of strokes, also disrupt the flow of blood to the brain, causing dizziness and nystagmus, an involuntary jerking of the eye. See Judith Hanson Lasater, *Yogabody: Anatomy, Kinesiology, and Asana* (Berkeley: Rodmell Press, 2009), pp. 51–52.

114 *describe the final journey:* Francesco Cacciola, Umesh Phalke, and Atul Goel, "Vertebral Artery in Relationship to C1-C2 Vertebrae: An Anatomical Study," *Neurology India*, vol. 52, no. 2 (June 2004), pp. 178–84.

114 *feeds such structures:* Adel K. Afifi and Ronald A. Bergman, *Functional Neuroanatomy: Text and Atlas*, 2nd ed. (New York: Lange Medical Books, 2005), pp. 352–54.

114 *recover most functions:* Abdullah Bin Saeed, Ashfaq Shuaib, Ghanem Al-Sulaiti, et al., "Vertebral Artery Dissection: Warning Symptoms, Clinical Features and Prognosis in 26 Patients," *Canadian Journal of Neurological Sciences*, vol. 27, no. 4 (November 2000), pp. 292–96; Wouter I. Schievink, "Spontaneous Dissection of the Carotid and Vertebral Arteries," *New England Journal of Medicine*, vol. 344, no. 12 (March 22, 2001), pp. 898–906.

114 *a prominent type:* Elisabeth Rosenthal, "Rare Threat of Stroke at the Beauty Salon," *New York Times*, April 28, 1993, Section C, p. 11.

115 *"as far back as possible":* Iyengar, *Light on Yoga*, pp. 107–109.

116 *"The body should be"*: Ibid., p. 211.

116 *a gruesome case study:* Willibald Nagler, "Vertebral Artery Obstruction by Hyperextension of the Neck: Report of Three Cases," *Archives of Physical Medicine and Rehabilitation*, vol. 54, no. 5 (May 1973), pp. 237–40; W. Nagler, "Mechanical Obstruction of Vertebral Arteries During Hyperextension of Neck," *British Journal of Sports Medicine*, vol. 7, nos. 1–2 (1973), pp. 92–97.

117 *An intermediate stage:* Iyengar, *Light on Yoga*, p. 358.

118 *according to a team in Chicago:* Steven H. Hanus, Terri D. Homer, and Donald H. Harter, "Vertebral Artery Occlusion Complicating Yoga Exercises," *Archives of Neurology*, vol. 34, no. 9 (1977), pp. 574–75.

118 *fascinated by the case:* Interviews, Steven H. Hanus, June 9 and July 1, 2011.

120 *medical team at the University of Hong Kong:* K. Y. Fong, R. T. Cheung, Y. L. Yu, et al., "Basilar Artery Occlusion Following Yoga Exercise: A Case Report," *Clinical and Experimental Neurology*, vol. 30 (1993), pp. 104–9; see also Robin, *A Physiological Handbook*, p. 516.

121 *a common feature of medical concern:* See, for example, W. Pryse-Phillips, "Infarction of the medulla and cervical cord after fitness exercises," *Stroke*, vol. 20 (1989), pp. 292–94; Daniel J. DeBehnke and William Brady, "Vertebral Artery Dissection Due to Minor Neck Trauma," *Journal of Emergency Medicine*, vol. 12, no. 1 (1994), pp. 27–31; Schievink, "Spontaneous Dissection."

121 Science of Flexibility: Michael J. Alter, *Science of Flexibility*, 3rd ed. (Champaign, IL: Human Kinetics, 2004), pp. 198–200.

121 *its surveys showed:* Letter to author, Vicky B. Leonard, Technical Information Specialist, U.S. Consumer Product Safety Commission, March 20, 2008. The enclosed data sheets from Leonard showed the results of a search I had requested of the National Electronic Injury Surveillance System for records citing the term "yoga" for the years 1996 through 2006.

122 *An analysis of the information:* By the author.

122–23 *found much of it fawning:* Robert Love, "Fear of Yoga," *Columbia Journalism Review*, November–December 2006, pp. 80–90.

123 *Stories appeared:* For example, see Leslie Kaminoff, with Coeli Carr, "Mr. Fix It for Injured Yoga Enthusiasts," *New York Times*, Aug 11, 2002, Section 3, p. 13; Stacey Colino, "The Wounded Warrior," *Washington Post*, April 16, 2002, Section HE, p. 1.

123 *The rising public debate:* During this period, the number of articles and scientific reports on yoga injury and safety approached two hundred. See Trisha Lamb, "Injuries from Yoga and Contraindications," a bibliography by the International Association of Yoga Therapists, April 27, 2006.

123 Body & Soul *magazine recounted:* Alanna Fincke, "Yoga Now: Bent Out of Shape," *Body & Soul,* March–April 2003, p. 40.

123 *An article in* The New York Times: Lorraine Kreahling, "When Does Flexible Become Harmful? 'Hot' Yoga Draws Fire," *New York Times,* March 30, 2004, Section F, p. 5.

124 *"twist and stretch with less chance":* Choudhury, *Bikram Yoga,* p. 74.

124 *told of being filmed one day:* Carol Krucoff, "Insight from Injury," *Yoga Journal,* May–June 2003, pp. 120–24, 203.

125 *"Yogi beware":* Judith Lasater, "Yogi Beware: Hidden Dangers Can Lurk Within Even the Most Familiar Pose," *Yoga Journal,* January–February 2005, pp. 110–19.

125 *told of how she had reinjured:* Kaitlin Quistgaard, "Safety Dance," *Yoga Journal,* March 2008, p. 12.

125 *began to run a legal proviso:* See, for instance, Anonymous, "Letters," *Yoga Journal,* February 2009, p. 16.

126 *"Proceed with Caution":* Catherine Guthrie, "Proceed with Caution: A Medical Review Points to the Potential Dangers of Sudden Neck Movements in Certain Poses," *Yoga Journal,* December 2001, p. 33.

127 *no scientist had ever published:* The studies lumped all known causes together. See Schievink, "Spontaneous Dissection"; Ralf W. Baumgartner, Julien Bogousslavsky, Valeria Caso, et al., eds., *Handbook on Cerebral Artery Dissection* (Basel: Karger Publishers, 2005), pp. 12–29, 44–53; Kwan-Woong Park, Jong-Sun Park, Sun-Chul Hwang, et al., "Vertebral Artery Dissection: Natural History, Clinical Features and Therapeutic Considerations," *Journal of Korean Neurosurgical Society,* vol. 44, no. 3 (2008), pp. 109–15.

128 *no general notice on the Internet:* Author's Google search, January 5, 2011.

128 *the Internet buzzed:* www.yogaforums.com/forums/f18/yoga-and-stroke-604.html.

129 *states began their regulatory effort:* A. G. Sulzberger, "Yoga Faces Regulation, and Firmly Pushes Back," *New York Times,* July 11, 2009, Section A, p. 1.

129 *involved a woman of twenty-nine:* Derek B. Johnson, Mathew J. Tierney, and Parvis J. Sadighi, "Kapalabhati Pranayama: Breath of Fire or Cause of Pneumothorax?" *Chest,* vol. 125, no. 5 (May 2004), pp. 1951–52.

129 *an emergency procedure:* Polly E. Parsons and John E. Heffner, eds., *Pulmonary/Respiratory Therapy Secrets* (Philadelphia: Hanley & Belfus, 2001), pp. 68–69.

130 *a joint letter:* Deane Hillsman and Vijai Sharma, "Yoga and Pneumothorax," *Chest,* vol. 127, no. 5 (May 2005), p. 1863.

130 *took to the pages:* Vijai P. Sharma, *"Pranayama* Can Be Practiced Safely," *International Journal of Yoga Therapy,* no. 17 (2007), pp. 75–79.

131 *they wrote in their 2007 report:* P. K. Sethi, A. Batra, N. K. Sethi, et al., "Compressive Cervical Myelopathy Due to Sirsasana, a Yoga Posture: A Case Report," *Internet Journal of Neurology,* vol. 6, no. 1 (2007).

131 *Even Iyengar got involved:* At his institute in Pune, Iyengar admitted that some postures in *Light on Yoga* might threaten injury. He called it a "dead book" and, to his credit, took an active role in pose redesign. For instance, students of Iyengar tell me that he pioneered the use of folded blankets to ease neck strain in Shoulder Stand, introducing the precaution as long ago as 1975. For Iyengar being quoted on the dated nature of his book, see Pat Musburger, "A Note from the President: Teachers and Teaching," Iyengar Yoga Association of the Northwest, *IYANW Update,* November 2007, p. 1. For his speaking openly of injuries and how to avoid them, see B.K.S. Iyengar, *Yoga: The Path to Holistic Health* (New York: Dorling Kindersley, 2007), pp. 7, 25, 194, 230, 243, 408.

132 *"The whole weight":* Iyengar, *Light on Yoga,* p. 187.

132 *called for exactly the reverse:* Richard Rosen, "Taking the Danger out of the Headstand," *Yoga World,* vol. 1, no. 9 (April–June 1999), pp. 3–4.

132 *"At this point in the game":* Yoga class with Mel Robin, Yoga Loft, Bethlehem, Pennsylvania, April 14, 2007.

132 *too dangerous for general yoga classes:* Timothy McCall, *Yoga as Medicine: The Yogic Prescription for Health and Healing* (New York: Bantam, 2007), pp. 499–500; Timothy McCall, "Upside Downside?" *Yoga Journal,* September–October 2003, p. 34.

132 *led to a condition:* McCall, *Yoga as Medicine,* pp. 89–90.

133 *"Knees are hinge joints":* Quoted in Martica K. Heaner, "Yoga's Softer Side," *Health,* vol. 15, no. 6 (July–August 2001), pp. 122–27. For a yoga anatomist who disagrees with the hinge metaphor, see Lasater, *Yogabody,* pp. 115–16.

133 *One of the most prolific:* For a short profile, see Anne Cushman, "Science Studies Yoga," *Yoga Journal*, August 1994, p. 43.

133 *spoken on yoga safety:* Anonymous, "Yoga Should Heal, Not Hurt, Says ACSM Expert," news release, American College of Sports Medicine, April 1, 2005.

133 *In one column:* Roger Cole, "Keep the Neck Healthy in Shoulderstand," *Yoga Journal*, www.yogajournal.com/for_teachers/1091.

133 *researchers in Europe:* Jani Mikkonen, Palle Pedersen, and Peter William McCarthy, "A Survey of Musculoskeletal Injury among *Ashtanga Vinyasa* Yoga Practitioners," *International Journal of Yoga Therapy*, no. 18 (2008), pp. 59–64.

134 *published a far more ambitious survey:* Loren M. Fishman, Ellen Saltonstall, and Susan Genis, "Understanding and Preventing Yoga Injuries," *International Journal of Yoga Therapy*, no. 19 (2009), pp. 1–8.

135 *trains its instructors for nine weeks:* For a portrait of Bikram training, and a glimpse of compassion, see Jeanne Heaton, "Teacher: Experience Needed," *New York Times*, March 13, 2011, Style section, p. 2.

135 *"It was a success":* Email, Glenn Black, September 30, 2009.

135 *caught up with him:* Interview, Glenn Black, Plaza Athénée, October 9, 2009.

V: Healing

137 *kept wanting to learn:* Interviews and emails, Loren M. Fishman, Manhattan, August and September, 2007, February and March, 2008.

137 *knocked on Iyengar's door:* The story is recounted in Loren Fishman and Carol Ardman, *Relief Is in the Stretch: End Back Pain Through Yoga* (New York: Norton, 2005), p. 5.

138 *a large shelf of Fishman's books:* In chronological order, they are Loren Fishman and Carol Ardman, *Back Talk: How to Diagnose and Cure Low Back Pain and Sciatica* (New York: Norton, 1997); Fishman and Ardman, *Relief*; Fishman and Ardman, *Sciatica Solutions: Diagnosis, Treatment, and Cure of Spinal and Piriformis Problems* (New York: Norton, 2006); Fishman and Ardman, *Cure Back Pain with Yoga* (New York: Norton, 2006); Fishman and Eric L. Small, *Yoga and Multiple Sclerosis: A Journey to Health and Healing* (New York: Demos Medical Publishing, 2007).

139 *his explanation for what stretching did:* Fishman and Ardman, *Relief*, pp. 18–21, 72.

140 *told of his therapeutic work:* Interview, Loren Fishman, August 10, 2007.

141 *a key feature:* Iyengar, *Light on Yoga*, p. 187.

142 *chronic pain in her right shoulder:* Nora Isaacs, "The Yoga Therapist Will See You Now," *New York Times*, May 10, 2007, Section G, p. 10.

142 *published them so other health professionals:* Loren M. Fishman and Caroline Konnoth, "Role of Headstand in the Management of Rotator Cuff Syndrome," *American Journal of Physical Medicine and Rehabilitation*, vol. 83, no. 3 (March 2004), p. 228; Loren M. Fishman, Caroline Konnoth, Alena Polesin, "Headstand for Rotator Cuff Tear: Shirshasana or Surgery," *International Journal of Yoga Therapy*, no. 16 (2006), pp. 39–47; Loren M. Fishman, Allen N. Wilkins, Tova Ovadia, et al., "Yoga-Based Maneuver Effectively Treats Rotator Cuff Syndrome," *Topics in Geriatric Rehabilitation*, vol. 27, no. 2 (2011), pp. 151–61. For a popular review, see Jane E. Brody, "Ancient Moves for Orthopedic Problems," *New York Times*, August 2, 2011, Section D, p. 7.

143 *had written a book:* Loren Fishman and Ellen Saltonstall, *Yoga for Arthritis: The Complete Guide* (New York: Norton, 2008).

144 *only rough outlines:* For Fishman's detailed prescription for treating spinal stenosis, see Fishman and Ardman, *Relief*, pp. 134–46.

144 *Fishman came in:* Yoga class with Loren M. Fishman, Manhattan Physical Medicine, April 8, 2008.

144 *showed us a simple treatment:* For a more detailed description, see Fishman and Saltonstall, *Yoga for Arthritis*, pp. 244–45.

145 *promoting a better quality of life:* Fishman and Small, *Yoga and Multiple Sclerosis*, pp. 2, 112–26.

147 *a device worn inside the vagina:* Medical regulators have questioned its safety after reports of injury and death. See Shirley S. Wang, "FDA Panel Takes Second Look," *Wall Street Journal*, September 8, 2011, Section B, p. 6.

150 *regularly uses the phrase:* See, for instance, Carol Krucoff, "Facing Cancer with Courage," *Yoga Journal*, December 2004, p. 143.

150 *"sensitive to individual needs":* Anonymous, "New Yoga Classes, Beginners to Advanced," *Record*, Montgomery County Recreation Department, Silver Spring, Maryland, vol. 1, no. 6 (March 2006).

150 *"30 years' experience":* http://search.barnesandnoble.com/Yoga-and-the
-Wisdom-of-Menopause/Suza-Francina/e/9780757300653.

151 *face no requirements:* Michael H. Cohen, "The Search for Regulatory Recognition of Yoga Therapy: Legal and Policy Issues," *International Journal of
Yoga Therapy*, no. 17 (2007), pp. 43–50.

151 *Yoga Alliance:* www.yogaalliance.org.

151 *"A growing number":* Georg Feuerstein, "Editorial," *International Journal
of Yoga Therapy*, no. 12 (2002), pp. 3–4.

151 *used liberally and often interchangeably:* Georg Feuerstein, "Editorial,"
International Journal of Yoga Therapy, no. 13 (2003), pp. 3–6.

152 *Anyone can claim:* For a discussion, see Lynn Somerstein, "Licensing is No
Merry-Go-Round," *International Journal of Yoga Therapy*, no. 18 (2008),
pp. 15–16.

152 *"provides in-depth training":* Anonymous, "Yoga Therapist Training," Namaste Institute for Holistic Studies, www.namasteinstitute.com/yogathera
pist.html.

152 *"There is no such thing":* Email to author, John Kepner, International Association of Yoga Therapists, May 23, 2008.

153 *its membership rolls increase:* Isaacs, "The Yoga Therapist."

154 *released a market study:* PR Newswire, "Yoga Journal Releases 2008 'Yoga
in America' Market Study," February 26, 2008, www.yogajournal.com/ad
vertise/press_releases/10.

154 *a good life:* A biographical sketch of Payne can be found in Larry Payne and
Richard Usatine, *Yoga Rx: A Step-by-Step Program to Promote Health,
Wellness, and Healing for Common Ailments* (New York: Broadway Books,
2002), pp. xiii–xiv.

155 *Payne was hooked:* Larry Payne, "President's Message," *Journal of the International Association of Yoga Therapists*, vol. 2, no. 1 (1991), p. v.

156 *Payne found a book:* Email to author, Larry Payne, Samata International,
August 13, 2011.

156 *federal investigators:* Robert J. Cramer, *Diploma Mills Are Easily Created
and Some Have Issued Bogus Degrees to Federal Employees at Government Expense* (Washington: Government Accountability Office, GAO-04
-1096T, September 23, 2004), pp. 3–4.

156 *blacklisted its degrees:* Oregon, for instance, ruled the professional use of degrees from Pacific Western as a misdemeanor carrying a punishment of up to six months in prison. See Oregon Student Assistance Commission, Office of Degree Authorization, "Use of unaccredited degrees in Oregon," www.osac.state.or.us/oda/unaccredited.aspx.

156 *when Payne got his doctorate:* Interview, Alan Baker, PWU Services, the official custodian of records for Pacific Western University, March 13, 2008. The Los Angeles school closed its doors in 2005.

156 *On his résumé:* Anonymous, "Larry Payne, Ph.D., Biographical Information," Samata International, www.samata.com/LarryP-4.php.

156 *"Look within yourself":* Home page on the Internet of Pacific Western University, August 3, 2002, as archived at web.archive.org/web/*/http://www.pwu-ca.edu.

156 *worked hard in his new job:* Larry Payne, "President's Message," *Journal of the International Association of Yoga Therapists,* vol. 1, nos. 1 and 2 (1990), p. v.

157 *"responds to his clients'":* Georg Feuerstein and Larry Payne, *Yoga for Dummies* (Indianapolis: Wiley, 1999), p. 2.

157 *"loads your blood":* Ibid., p. 64.

157 *"allows you to take in":* Ibid., p. 67.

157 *"treat your body":* Ibid., p. 71.

158 *"steps up your metabolism":* Ibid., p. 67.

158 *"boost your metabolism":* Ibid., p. 281.

158 *"helps you step up":* Ibid., p. 331.

158 *"the best manager":* Ibid., p. 67.

158 *"keep the rolls":* Ibid., p. 331.

158 *"helps you shed":* Ibid., p. 344.

159 *"Health Maintenance and Restoration":* Ibid., pp. 87–228.

159 *a Samata news release:* Chris Fletcher, "Larry Payne, Ph.D., Selected as First Yoga Teacher to Participate in the World Economic Forum," Samata Yoga Center, May 2000.

159 *a close relationship:* Payne and Usatine, *Yoga Rx,* p. xiv.

160 *"on every page":* Ibid., p. xviii.

160 *devoted a long section:* Ibid., pp. 249–63.

160 *Weight Management for People:* Larry Payne, *Larry Payne's Yoga Rx Therapy: Weight Management for People with Curves*, DVD, Newport Media & Samata International, 2005.

160 *he helped found:* John Weeks, "Yoga Rx University Yoga Certificate: Yoga Therapy's Future?" *Yoga Therapy in Practice*, December 2006, pp. 28–29; Anonymous, "Larry Payne Establishes New Yoga Rx Therapy Certification Program at Loyola Marymount University," *Healthy News Service*, August 17, 2005.

161 *Payne told me:* Emails, Larry Payne, Samata International, June 15, 2011, and August 13, 2011.

161 *"Yoga is in danger":* Interview, Loren M. Fishman, Manhattan Physical Medicine, February 7, 2008.

VI: Divine Sex

163 *The Complete Illustrated Book:* Vishnudevananda, *The Complete Illustrated Book.*

163 *One said yoga reduced:* T. Schmidt, A. Wijga, A. Von Zur Mühle, et al., "Changes in Cardiovascular Risk Factors and Hormones During a Comprehensive Residential Three Month Kriya Yoga Training and Vegetarian Nutrition," *Acta Physiologica Scandinavica, Supplement*, vol. 640 (1997), pp. 158–62.

163 *speaking with great candor:* See, for instance, Yogani, *Advanced Yoga Practices: Easy Lessons for Ecstatic Living* (Nashville: AYP Publishing, 2004); and Stuart Sovatsky, "On Being Moved," in Tami Simon, ed., *Kundalini Rising: Exploring the Energy of Awakening* (Boulder, CO: Sounds True, 2009), pp. 247–67.

163 *called her blinding ecstasies:* Interviews, Jana Dixon, author of *The Biology of Kundalini: Exploring the Fire of Life* (Raleigh, NC: Lulu.com, 2008), July 23 and 24, 2009.

164 *"harder, deeper, and further":* Jacquie Noelle, *Better Sex Through Yoga, Advanced* (Los Angeles: Starlight Home Entertainment, 2003).

164 *the world's most celebrated gurus:* For an overview, see Geoffrey D. Falk, *Stripping the Gurus: Sex, Violence, Abuse and Enlightenment* (Toronto: Million Monkeys Press, 2009).

164 *victims who tended to rationalize:* The women have spoken out only occasionally, and often anonymously, but have said enough in public to give

a sense of their views. For a glimpse, see the website Leaving Siddha Yoga, www.leavingsiddhayoga.net, as well as the Adi Da Archives, www.adidaar chives.org. Joan Bridges was one of the first women to speak of her experiences without the shield of anonymity. See her website, www.shadowofthe guru.com

164 *protestors waving placards:* Rex Springston, "Yoga Guru Devotees Rally to His Support," *Richmond Times-Dispatch*, August 3, 1991, p. 13.

164 *the man who impressed scientists:* The swami also had a reputation for philandering. See Katharine Webster, "The Case Against Swami Rama of the Himalayas," *Yoga Journal*, November–December 1990, p. 58.

164 *evaded deposition:* Memorandum, Thomas I. Vanaskie, chief judge, Middle District of Pennsylvania, "Jasmine Patel versus Himalayan International Institute of Yoga Science and Philosophy of the USA," December 9, 1999, p. 12.

165 *Pennsylvania jury awarded:* Ken Dilanian, "$1.9 Million Awarded in Sex Scandal," *Philadelphia Inquirer*, September 6, 1997, p. B1.

165 *confessed to multiple affairs:* Jon Auerbach, "Yoga Guru Who Taught Virtue of Celibacy Admits to Affairs," *Boston Globe*, November 2, 1994, p. 22; William A. Davis, "A Guru's Fall from Grace," *Boston Globe*, December 22, 1994, p. 61; also see Anonymous, "History of Kripalu Center," Kripalu Center for Yoga and Health, undated, www.kripalu.org/about_us/491/.

165 *asked about rumors:* Clancy Martin, "The Overheated, Oversexed Cult of Bikram Choudhury," *Details*, February 2011, p. 92.

166 *When Abraham Morgentaler wrote:* Abraham Morgentaler, *Testosterone for Life: Recharge Your Vitality, Sex Drive, Muscle Mass & Overall Health!* (New York: McGraw-Hill, 2009).

166 *Orgasm, Inc.:* Liz Canner, "Orgasm Inc.: The Strange Science of Female Pleasure," orgasminc.org.

166 *The formal diagnosis:* K. N. Udupa, *Stress and Its Management by Yoga* (Delhi: Motilal Banarsidass, 2000), p. vii.

167 *improve a patient's hormone profile:* Ibid., pp. 149–61.

167 *studied a dozen young men:* J. D. Gode, R. H. Singh, R. M. Settiwar, et al., "Increased Urinary Excretion of Testosterone Following a Course of Yoga in Normal Young Volunteers," *Indian Journal of Medical Sciences*, vol. 28, nos. 4–5 (April–May 1974), pp. 212–15.

168 *"considerable improvement":* Udupa, *Stress*, p. 154.

169 *"vitality and sexual vigour"*: Swami Shankardevananda Saraswati, "Evaluating Yoga Research," *Yoga Magazine*, June 1979, www.yogamag.net/archives/1979/fjune79/evalre.shtml.

169 *acts to improve mood:* K. Christiansen, "Behavioral Correlates of Testosterone," in Eberhard Nieschlag and Hermann M. Behre, eds., *Testosterone: Action, Deficiency, Substitution,* 3rd ed. (Cambridge: Cambridge University Press, 2004), pp. 142–46.

169 *shown to bolster attention:* Ibid., pp. 146–50.

169 *important role in female arousal:* S. Bolour and G. Braunstein, "Testosterone Therapy in Women: A Review," *International Journal of Impotence Research,* vol. 17 (2005), pp. 399–408.

169 *studies have linked testosterone:* Christiansen, "Behavioral Correlates," in Nieschlag and Behre, *Testosterone,* pp. 130–33.

170 *closely studying the hormone:* Annalee Newitz, "The Coming Boom," *Wired,* July 2005, pp. 106–10; Jane E. Brody, "A Libido Drug for Women?" *New York Times,* March 31, 2009, Section D, p. 7; Jane E. Brody, "A Dip in the Sex Drive, Tied to Menopause," *New York Times,* March 31, 2009, Section D, p. 7.

170 *scientists at the Hannover:* Schmidt et al., "Changes in Cardiovascular Risk Factors."

170 *vegetarianism alone reduces:* See, for instance, A. Raben, B. Kiens, E. A. Richter, et al., "Serum Sex Hormones and Endurance Performance After a Lacto-ovo Vegetarian and a Mixed Diet," *Medicine and Science in Sports and Exercise,* vol. 24, no. 11 (November 1992), pp. 1290–97.

171 *"You won't boost testosterone":* Al Sears, "Manhood Banned!" www.alsearsmd.com/pdf/manhoodbanned.pdf.

171 *"recharges your sex life":* John Capouya, *Real Men Do Yoga* (Deerfield Beach, FL: Health Communications, 2003), pp. xv, 170.

171 *A photo of Minvaleev:* Wikipedia, "Rinad Minvaleev," en.wikipedia.org/wiki/Rinad_Minvaleev.

171 *study with a very narrow focus:* R. S. Minvaleev, A. D. Nozdrachev, V. V. Kiryanova, et al., "Postural Influences on the Hormone Level in Healthy Subjects: I. The Cobra Posture and Steroid Hormones," *Human Physiology,* vol. 30, no. 4 (2004), pp. 452–56.

172 *a holy book of Hatha:* Mallinson, *Gheranda,* pp. xiv–xvi.

172 *"the physical fire"*: Ibid., p. 54.

172 *the concluding step:* Ibid., p. 72.

172 *"until the pubis"*: Iyengar, *Light on Yoga*, p. 107.

173 *Runners, for instance:* McArdle et al., *Exercise Physiology*, pp. 448, 450.

173 *Dostálek fell for yoga:* Letter to author, Ctibor Dostálek, July 6, 2009.

173 *named director: curriculum vitae,* Ctibor Dostálek, undated, p. 1.

174 *held the practice in such esteem:* Swami Rama, *Path of Fire and Light: Volume 2, A Practical Companion to Volume 1* (Honesdale, PA: Himalayan Institute Press, 1988), pp. 175–76.

174 *Doing Agni Sara properly:* See H. David Coulter, *Anatomy of Hatha Yoga: A Manual for Students, Teachers, and Practitioners* (Honesdale, PA: Body and Breath, 2001), pp. 188–200; the exercise has many variations, including the one that I describe.

174 *Masters and Johnson reported:* William H. Masters and Virginia E. Johnson, *Human Sexual Response* (Boston: Little, Brown, 1966), pp. 295–98.

174 *to a second brain:* Harriet Brown, "The Other Brain, the One with Butterflies, Also Deals with Many Woes," *New York Times*, August 23, 2005, Section F, p. 5; Michael D. Gershon, *The Second Brain* (New York: Harper-Collins, 1998).

174 *envelops the viscera:* Mark F. Bear, Barry W. Connors, and Michael A. Paradiso, *Neuroscience*, 3rd ed. (Philadelphia: Lippincott Williams & Wilkins, 2006), pp. 495–96.

175 *bursts of brain excitation:* E. Roldán and C. Dostálek, "Description of an EEG Pattern Evoked in Central-Parietal Areas by the Hathayogic Exercise Agnisara," *Activitas Nervosa Superior*, vol. 25, no. 4 (1983), pp. 241–46.

175 *"paroxysmal"*: E. Roldán and C. Dostálek, "EEG Patterns Suggestive of Shifted Levels of Excitation Effected by Hathayogic Exercises," *Activitas Nervosa Superior*, vol. 27, no. 2 (1985), pp. 81–88.

175 *did so in the pages:* Ctibor Dostálek, "YOGA: A Returning Constituent of Medical Sciences," *Yoga Mimansa*, vol. 24, no. 2 (July 1985), pp. 21–34.

176 *A few scientists glimpsed:* See, for instance, N. N. Das and H. Gastaut, "Variations de l'activité électrique du cerveau, du coeur et des muscles au

cours de la méditation et de l'extase yogique," *Neurophysiologie Clinique*, supplement, vol. 6 (1957), pp. 211–19; M. A. Wenger and B. K. Bagchi, "Studies of Autonomic Functions in Practitioners of Yoga in India," *Behavioral Science*, vol. 6 (1961), pp. 312–23.

176 *did the most thorough study:* James C. Corby, Walton T. Roth, Vincent P. Zarcone, Jr., et al., "Psychophysiological Correlates of the Practice of Tantric Yoga Meditation," *Archives of General Psychiatry*, vol. 35, no. 5 (1978), pp. 571–77.

176 *important sign of emotional arousal:* The linkage between skin conductance and sexual arousal appears to be variable. See, for instance, Mary Lake Polan, John E. Desmond, Linda L. Banner, et al., "Female Sexual Arousal: A Behavioral Analysis," *Fertility and Sterility*, vol. 80, no. 6 (December 2003), pp. 1480–87.

176 *way to probe the unconscious:* L. Binswanger, "On the Psychogalvanic Phenomenon in Association Experiments," in C. G. Jung, *Studies in Word Association* (New York: Russell and Russell, 1969), pp. 446–530.

177 *equal to that of frenzied lovers:* Masters and Johnson, *Human Sexual Response*, pp. 34–35, 174, 278.

178 *reported rates of more than forty breaths:* Ibid., pp. 34, 277–78.

178 *Advanced students are encouraged:* Coulter, *Anatomy*, p. 119.

179 *wraps around the brain stem:* Afifi and Bergman, *Functional Neuroanatomy*, pp. 280–96.

179 *body made of two lobes:* Ibid., pp. 289–92.

179 *Dutch scientists recently studied:* Guido A. van Wingen, Stas A. Zylicz, and Sara Pieters, et al., "Testosterone Increases Amygdala Reactivity in Middle-Aged Women to a Young Adulthood Level," *Neuropsychopharmacology*, vol. 34 (2009), pp. 539–47.

179 *repercussions on different parts:* Stefan Posse, Uwe Olthoff, Matthias Weckesser, et al., "Regional Dynamic Signal Changes During Controlled Hyperventilation Assessed with Blood Oxygen Level–Dependent Functional MR Imaging," *American Journal of Neuroradiology*, vol. 18 (October 1997), pp. 1763–70; Janniko R. Georgiadis, Rudie Kortekaas, Rutger Kuipers et al., "Regional Cerebral Blood Flow Changes Associated with Clitorally Induced Orgasm in Healthy Women," *European Journal of Neuroscience*, vol. 24 (2006), pp. 3305–16.

180 *a theory of sex hyperventilation:* Torsten Passie, Uwe Hartmann, Udo Schneider, et al., "On the Function of Groaning and Hyperventilation

During Sexual Intercourse: Intensification of Sexual Experience by Altering Brain Metabolism Through Hypocapnia," *Medical Hypotheses*, vol. 60, no. 5 (May 2003), pp. 660–63.

180 *Lori A. Brotto and other sex researchers:* For a profile, see Daniel Bergner, "Women Who Want to Want," *New York Times Magazine*, November 29, 2009, Section MM, p. 42.

180 *rise in the amplitude:* Lori A. Brotto and Boris B. Gorzalka, "Genital and Subjective Sexual Arousal in Postmenopausal Women: Influence of Laboratory-Induced Hyperventilation," *Journal of Sex and Marital Therapy*, vol. 28 (2002), pp. 39–53.

181 *fast breathing could improve arousal:* Lori A. Brotto, Carolin Klein, and Boris B. Gorzalka, "Laboratory-Induced Hyperventilation Differentiates Female Sexual Arousal Disorder Subtypes," *Archives of Sexual Behavior*, vol. 38, no. 4 (August 2009), pp. 463–75.

181 *a routine of twenty-six poses:* Choudhury details the typical class in *Bikram Yoga*.

181 *muscular strain as an integral part:* Masters and Johnson, *Human Sexual Response*, pp. 294–300.

182 *puts the make on a nearby guy:* www.youtube.com/watch?v=3rKR4tRevfU &feature=related.

182 *campaign quizzed readers:* Associated Press, "Berra Sues Network Over Ad," *New York Times*, February 2, 2005, Section B, p. 6.

182 *"Not tonight, hon":* Michael Crawford, "Not Tonight, Hon," *New Yorker*, February 23, 2009, p. 50.

182 *colleagues published two papers:* Vikas Dhikav, Girish Karmarkar, Richa Gupta, et al., "Yoga in Female Sexual Functions," *Journal of Sexual Medicine*, vol. 7, no. 2, part 2 (February 2010), pp. 964–70; Vikas Dhikav, Girish Karmarkar, Myank Verma, et al., "Yoga in Male Sexual Functioning: A Noncompararive Pilot Study," *Journal of Sexual Medicine*, vol. 7, no. 10 (October 2010), pp. 3460–66.

183 *a few seconds and twenty-two seconds:* Barry R. Komisaruk, Beverly Whipple, Sara Nasserzadeh, et al., *The Orgasm Answer Guide* (Baltimore: Johns Hopkins University Press, 2009), pp. 9–10.

183 *orgasms that lasted a minute or more:* Masters and Johnson, *Human Sexual Response*, pp. 131–32.

183 *The most important shift featured:* Bear et al., *Neuroscience*, pp. 542–43; Otto Appenzeller and Emilio Oribe, *The Autonomic Nervous System*, 5th ed. (New York: Elsevier, 1997), pp. 339–52.

184 *a remarkable class of women:* Beverly Whipple, Gina Ogden, and Barry R. Komisaruk, "Physiological Correlates of Imagery-Induced Orgasm in Women," *Archives of Sexual Behavior*, vol. 21, no. 2 (April 1992), pp. 121–33.

184 *yoga played a central role:* The Rutgers scientists made no mention of the yoga connection in their paper. But one of them, Gina Ogden, discussed the link in a book she subsequently wrote, *Women Who Love Sex: Ordinary Women Describe Their Paths to Pleasure, Intimacy, and Ecstasy* (Boston: Trumpeter, 2007), pp. 111–37.

184 *"Just tell me":* Quoted in ibid., p. 114.

185 *Christian ascetics also evoked:* For a scholarly discussion, see Chuck M. MacKnee, "Peak Sexual and Spiritual Experience: Exploring the Mystical Relationship," *Theology and Sexuality*, vol. 3 (1996), pp. 97–115.

185 *kept assistants on hand:* Swami Chetanananda, *Ramakrishna as We Saw Him* (St. Louis: Vedanta Society of St. Louis, 1990), pp. 154–55.

186 *the snake has a long history:* Eliade, *Yoga*, p. 165.

186 *"like a snake":* Swami Nikhilánanda, trans., ed., *The Gospel of Sri Ramakrishna* (New York: Ramakrishna-Vivekananda Center, 1942), p. 830.

186 *kundalini as a "great fire":* Eliade, *Yoga.*, pp. 246–47, 330–34.

186 *means "to heat or burn":* Leza Lowitz and Reema Datta, *Sacred Sanskrit Words* (Berkeley: Stone Bridge Press, 2004), p. 111.

186 *the mystic fire as divine in origin:* Eliade, *Yoga*, pp. 49–50, 104, 200–73.

186 *vague in describing the physical basis:* Scholars have found a few ancient claims about the transmutation of sexual energy amid a wealth of euphemisms and fiery metaphors. See Ibid., *Yoga*, pp. 246, 331.

186 *definitions include mystic energy:* See Fernando Pagés Ruiz, "Too Hot to Handle? Stuart Sovatsky and Shanti Shanti Kaur Khalsa Discuss How to Kindle Kundalini Without Getting Burned," *Yoga Journal*, March–April 2002, pp. 161–64.

186 *rejects such portrayals:* Yogani, *Advanced*, pp. 69–70.

186 *"a flowering of orgasm":* Ibid., p. 415.

187 *a case of kundalini arousal:* C. G. Jung, *The Collected Works of C. G. Jung,* vol. 16, *The Practice of Psychotherapy,* R. F. C. Hull, trans. (Princeton, NJ: Princeton University Press, 1966), pp. 330–37; see also Sonu Shamdasani, ed., *The Psychology of Kundalini Yoga: Notes of the Seminar Given in 1932 by C. G. Jung* (Princeton, NJ: Princeton University Press, 1999), pp. 104–106.

187 *warned people away:* Ibid., pp. xxix–xxx.

187 *One of his sternest admonitions:* For the dating of the warning to 1938, see W. Y. Evans-Wentz, ed., *The Tibetan Book of the Dead* (New York: Oxford University Press, 2000), p. vii.

187 *"deliberately induced psychotic state":* C. G. Jung, "Psychological Commentary," in Evans-Wentz, *The Tibetan Book,* p. xlvi.

187 *the San Francisco psychiatrist:* For a biographical sketch, see Lee Sannella, *On Genius: An Evolutionary Force Inherent in Every Being* (West Conshohocken, PA: Infinity Publishing, 2006), pp. 174–82; for his tie to Esalen, see Jeffrey John Kripal, *Esalen: America and the Religion of No Religion* (Chicago: University of Chicago Press, 2007), p. 499.

188 *told of thirteen people:* Lee Sannella, *The Kundalini Experience: Psychosis or Transcendence?* (Lower Lake, CA: Integral Publishing, 1992). This is a later edition and the source for page citations.

188 *devoted one sentence:* Ibid., p. 7.

188 *his portrayal of the Reverend:* Ibid., pp. 36–37. While Sannella's portraits were anonymous, biographical details often gave away the identity. For the reverend's own account, see John Scudder, "A Psychic Healer Experiences Kundalini," in John White, ed., *Kundalini: Evolution and Enlightenment* (Saint Paul, MN: Paragon House, 1990), pp. 189–97.

189 *nearly one thousand cases:* Stanislav Grof and Christina Grof, "Spiritual Emergency: Understanding Evolutionary Crises," in Stanislav Grof and Christina Grof, eds., *Spiritual Emergency: When Personal Transformation Becomes a Crisis* (New York: Putnam, 1989), p. 15.

189 *more than five hundred calls:* Jeneane Prevatt and Russ Park, "The Spiritual Emergence Network (SEN)," in Grof and Grof, eds., *Spiritual Emergency,* p. 227.

189 *some tell of terrors:* See, for instance, the Swedish website "Kundalini Short Circuits—Risks & Information," www.kundalini.se/eng.

189 *tell of heart attacks:* See, for instance, Mystress Angelique Serpent, "Doctors," www.kundalini-teacher.com/symptoms/doctors.php, and "Kundalini

Awakened Through Grace: Writings by and about Rick Puravs," www.nonduality.com/puravs.htm.

189 *told of his own arousal:* Bob Boyd's website has disappeared but his autobiographical essay, "The Safety of the Heart," can be found at www.elcollie
.com/st/support.html.

189 *paints an alluring picture:* Elizabeth Gilbert, *Eat, Pray, Love: One Woman's
Search for Everything Across Italy, India and Indonesia* (New York: Penguin, 2006), pp. 141–46, 158–59, 197–200.

189 *More University:* For a profile, see K. L. Billingsley, "University of Sex,"
Heterodoxy, vol. 2, no. 7 (March 1994), p. 1.

189 *no library and no campus:* Anonymous, "California Trying to Close
Worthless-Diploma Schools," *New York Times*, August 31, 1994, Section
B, p. 8.

190 *kept going for eleven hours:* Leah Schwartz and Bob Schwartz, *The One-
Hour Orgasm* (New York: St. Martin's, 2006), p. 3.

190 *"I was breathing fire":* Interview, Patricia Taylor, February 25, 2010.

190 *she authored* Expanded Orgasm*:* Patricia Taylor, *Expanded Orgasm: Soar to
Ecstasy at Your Lover's Every Touch* (Naperville, IL: Sourcebooks, 2002).

190 *more than one hundred papers: curriculum vitae,* Barry R. Komisaruk,
March 2010.

190 *an understated book:* Barry R. Komisaruk, Carlos Beyer-Flores, and Beverly
Whipple, *The Science of Orgasm* (Baltimore: Johns Hopkins University
Press, 2006).

191 *began a new round of experimentation:* Email to author, Barry Komisaruk,
April 12, 2010.

191 *Komisaruk was attracted:* Interview, Barry Komisaruk, Rutgers University,
April 8, 2010.

192 *"It's the least sexy thing":* Interview, Nan Wise, Rutgers University, April 3,
2010.

192 *One group she drew on:* For a portrait, see Patricia Leigh Brown and Carol
Pogash, "The Pleasure Principle," *New York Times*, March 15, 2009, Style
section, p. 8.

VII: Muse

195 *Paul Pond wanted to know:* This sketch of Pond and the founding of the Institute for Consciousness Research are based on interviews conducted in Canada with Paul Pond, Teri Degler, and Michael Bradford, all founding ICR members, August 1 and 2, 2009.

195 *published in* Physical Review: Paul Pond, "Hard-Meson Calculation of Kπ Scattering," *Physical Review D*, vol. 3, no. 9 (1971), pp. 2195–209.

195 *"a virtuoso of a high order":* Gopi Krishna, *The Biological Basis of Religion and Genius* (New York: Harper & Row, 1972), p. 98.

196 *examined such figures as Brahms:* Institute for Consciousness Research, "Research and Articles," www.icrcanada.org/research.html.

196 *underwent his own transformation:* Teri Degler, *Fiery Muse: Creativity and the Spiritual Quest* (Toronto: Random House, 1996), pp. 98–100.

196 *Restless ego:* Paul Pond, "The Road Home," *Journal of Religion and Psychical Research*, vol. 16 (January 1993), p. 41.

196 *"Before the problem":* Sigmund Freud, "Dostoevsky and Parricide," in James Strachey and Anna Freud, eds., *The Standard Edition of the Complete Psychological Works of Sigmund Freud*, vol. 21, *The Future of an Illusion, Civilization and Its Discontents and Other Works* (London: Hogarth Press, 1953), p. 177.

197 *turned to the discipline relatively early:* C. G. Jung, Aniela Jaffé, ed., Richard and Clara Winston, trans., *Memories, Dreams, Reflections* (New York: Vintage, 1989), pp. 170–77.

197 *hailed as the genesis:* Sara Corbett, "The Holy Grail of the Unconscious," *New York Times Magazine*, September 20, 2009, Section MM, p. 34.

197 *"I was frequently":* Jung and Jaffé, *Memories*, p. 177.

198 *became a confirmed health enthusiast:* William Ander Smith, *The Mystery of Leopold Stokowski* (Madison, NJ: Fairleigh Dickinson University Press, 1990), p. 105.

198 *would meditate to clear his mind:* William A. Smith, "Leopold Stokowski: A Re-Evaluation," *American Music*, vol. 1, no. 3 (Autumn 1983), pp. 23–37.

198 *taught her yoga:* Frederick Sands and Sven Broman, *The Divine Garbo* (New York: Grosset and Dunlap, 1979), pp. 188–91; Antoni Gronowicz, *Garbo: Her Story* (New York: Simon & Schuster, 1990), p. 353.

198 *recounted how Garbo taught him:* Gayelord Hauser, *Gayelord Hauser's Treasury of Secrets* (New York: Fawcett World Library, 1967), p. 198.

198 *performed hundreds of times:* Hunphrey Burton, *Yehudi Menuhin: A Life* (Boston: Northeastern University Press, 2001), pp. 223–53.

198 *a courageous act of reconciliation:* Ibid., pp. 282–86.

199 *he met Iyengar:* Ibid., pp. 331–32; Iyengar, *Iyengar*, pp. 59–64.

199 *wrote a foreword of considerable grace:* Yehudi Menuhin, "Foreword," in Iyengar, *Light on Yoga*, pp. 11–12.

199 *told an interviewer that it can:* Ganga White, "Every Breath You Take: Sting on Yoga," *Yoga Journal*, November–December 1995, pp. 64–69. For more on the musician's views about yoga, see Sting, "Foreword: The Yogi and the Shower Singer," in Ganga White, *Yoga Beyond Belief: Insights to Awaken and Deepen Your Practice* (Berkeley: North Atlantic Books, 2007), pp. xiii–xvi.

199 *A cottage industry has sprung up:* Debra Bokur, "Spiritual Retreats: The Inside Story," *Yoga Journal*, December 2003, pp. 46–48.

200 *"Yoga won't make writing easy":* Quoted in Anonymous, "The Next Wave of Yoga Research: Creativity?" October 19, 2007, www.prleap.com /pr/98937.

200 *Novick's book:* Linda Novick, *The Painting Path: Embodying Spiritual Discovery through Yoga, Brush and Color* (Woodstock, VT: SkyLight Paths, 2007).

200 *"The students," she recalled:* Mia Olson, *Musician's Yoga* (Boston: Berklee Press, 2009), p. 125.

200 *not unusual for a beginning student:* Robin, *A Physiological Handbook*, p. 150.

200 *bursts of long-suppressed emotion:* It turns out that many practices that seek to promote serenity—meditation, yoga, massage, to name a few—can spark emotional flare-ups. See Amy Eden Jollymore, "Emotional Ambush," *Natural Health*, November–December 1999, pp. 87–89.

201 *examined the roots of creative reverie:* Green and Green, *Beyond Biofeedback*, pp. 118–52, 255–56.

202 *"a quick way to calm":* Jeff Davis, *The Journey from the Center to the Page: Yoga Philosophies and Practices as Muse for Authentic Writing* (Rhinebeck, NY: Monkfish Book Publishing, 2008), p. 41.

202 *details the favorite drinks:* Mark Bailey, *Hemingway & Bailey's Bartending Guide to Great American Writers* (Chapel Hill, NC: Algonquin Books, 2006).

202 *does the trick indirectly:* Bear et al., *Neuroscience,* pp. 156–57, 670–71.

203 *the investigations of Roger Sperry:* Stanley Finger, *Minds Behind the Brain: A History of the Pioneers and Their Discoveries* (New York: Oxford University Press, 2005), pp. 281–300. For a profile of one of Sperry's students and his role in the discoveries, see Benedict Carey, "Michael S. Gazzaniga: Decoding the Brain's Cacophony," *New York Times,* November 1, 2011, Section D, p. 1.

203 *basic difference between the two halves:* Jill Bolte Taylor, *My Stroke of Insight: A Brain Scientist's Personal Journey* (New York: Viking, 2006), pp. 27–36, 137–45.

203 *inconspicuous type of sensory activity:* Faith Hickman Brynie, *Brain Sense: The Science of the Senses and How We Process the World Around Us* (New York: AMACOM, 2009), pp. 18–19.

204 *Jill Bolte Taylor:* For a profile, see Leslie Kaufman, "A Superhighway to Bliss," *New York Times,* May 25, 2008, Style section, p. 1.

205 *"I felt like a genie":* Taylor, *My Stroke,* p. 67.

205 *learning how to empower:* Ibid., pp. 159–74.

205 *In 2001, he and his colleagues reported:* Andrew B. Newberg, Abass Alavi, Michael J. Baime, et al., "The Measurement of Regional Cerebral Blood Flow During the Complex Cognitive Task of Meditation: A Preliminary SPECT Study," *Psychiatry Research: Neuroimaging,* vol. 106, no. 2 (April 2001), pp. 113–22.

205 *a more detailed portrait in 2007:* Andrew Newberg, Mark Waldman, Nancy Wintering, et al., "Cerebral blood flow effects in long-term meditators," *Journal of Nuclear Medicine,* vol. 48, supplement 2 (2007), p. 111P.

206 *"We found greater overall activations":* Debbie L. Cohen, Nancy Wintering, Victoria Tolles, et al., "Cerebral Blood Flow Effects of Yoga Training: Preliminary Evaluation of 4 Cases," *Journal of Alternative and Complementary Medicine,* vol. 15, no. 1 (2009), pp. 9–14.

206 *suggests that the right hemisphere orchestrates:* George J. Demakis, "Sex and the Brain," in Richard D. McAnulty and M. Michele Burnette, eds., *Sex and Sexuality,* vol. 2, *Sexual Function and Dysfunction* (Westport, CT: Praeger, 2006), pp. 28–34.

NOTES

207 *linked sexual excitement to the lighting:* See, for instance, Jari Tiihonen, Jyrki Kuikka, Jukka Kupila, et al., "Increase in Cerebral Blood Flow of Right Prefrontal Cortex in Man During Orgasm," *Neuroscience Letters,* vol. 170, no. 2 (April 11, 1994), pp. 241–43.

207 *a study of four hundred and twenty-five:* Daniel Nettle and Helen Clegg, "Schizotypy, Creativity and Mating Success in Humans," *Proceedings of the Royal Society, Biological Sciences,* vol. 273, no. 1586 (March 7, 2006), pp. 611–15.

208 *An astonishing case:* Oliver Sacks, *Musicophilia: Tales of Music and the Brain* (New York: Knopf, 2008), pp. 3–17.

209 *all smiles and applause:* The PBS science show *Nova* devoted a segment to Sacks's book and Cicoria. See "Musical Minds," www.pbs.org/wgbh/nova/musicminds/about.html.

209 *a CD of classical piano solos:* Tony Cicoria, *Notes from an Accidental Pianist and Composer,* www.cdbaby.com/cd/drtonycicoria.

209 *an unending flow of poetry:* Gopi Krishna, *Kundalini: The Evolutionary Energy in Man* (Boston: Shambhala, 1997), pp. 200–202, 206–13.

209 *"I had never learned German":* Ibid., p. 212.

210 *"It is, if one may say so":* Carl von Weizsäcker, "Introduction," in Krishna, *The Biological Basis,* pp. 20–21.

210 *a diverse body of artwork:* Adi Da up close, "Art and Photography," www.adidaupclose.org/Art_and_Photography/index.html.

210 *His 2007 book:* Adi Da Samraj (Franklin Jones), *The Spectra Suites* (New York: Welcome Books, 2007).

210 *a page devoted to her paintings:* Jana Dixon, "Artwork," biologyofkundalini.com/article.php?story=Artwork.

211 *"We're everyday people":* Interview, Dale Pond, August 1, 2009.

212 *profiled by Degler in a book:* Degler, *Fiery Muse,* pp. 44–50, 186.

213 *Beneath the surface:* Neil Bethell Sinclair, *The Spirit Flies Free: The Kundalini Poems* (Bayside, CA.: Life Force Books, 2008), p. 2.

Epilogue

215 *arrived at a turning point:* My approach here was inspired by Fishman and Saltonstall, "Authors' Note," *Yoga for Arthritis,* pp. 15–17.

217 *more than $2* trillion *a year:* Robert Pear, "Health Spending Exceeded Record $2 Trillion in 2006," *New York Times,* January 8, 2008, Section A, p. 20.

219 *In his book:* Dalai Lama, *The Universe in a Single Atom: The Convergence of Science and Spirituality* (New York: Morgan Road, 2005), pp. 3, 13.

219 *a new cycle of studies:* For a list of yoga studies that the NIH funds, enter the search term "yoga" at its Reporter site: www.projectreporter.nih.gov/re porter.cfm.

220 *ridiculed yoga studies:* Terence P. Jeffrey, "WASTE: Federal 'Gurus' Funding Yoga," *Human Events,* July 20, 2005, www.humanevents.com/article .php?id=8165.

220 *amounted to about $7 million:* Author search of the NIH Reporter site, www.projectreporter.nih.gov, October 31, 2011. I used the word *yoga* but— to focus on major efforts and eliminate ones in which the discipline played a minor role—limited the categories to projects in which the word appeared in project titles, project terms, and abstracts. In fiscal 2011, the result was 26 research projects that had a total funding of $6,563,721.

222 *devoted its last chapter:* William J. Broad, *The Oracle: The Lost Secrets and Hidden Message of Ancient Delphi* (New York: Penguin, 2006), pp. 227–50.

Bibliography

Sources cited multiple times appear below as well as in the Notes, whereas those cited once appear exclusively in the Notes.

Afifi, Adel K., and Ronald A. Bergman. *Functional Neuroanatomy: Text and Atlas*, 2nd ed. New York: Lange Medical Books, 2005.

Akers, Brian Dana, trans. *The Hatha Yoga Pradipika*. Woodstock, NY: YogaVidya.com, 2002.

Alter, Joseph S. *Gandhi's Body: Sex, Diet, and the Politics of Nationalism*. Philadelphia: University of Pennsylvania Press, 2000.

_____. *Yoga in Modern India: The Body Between Science and Philosophy*. Princeton, NJ: Princeton University Press, 2004.

Bauman, Alisa. "Is Yoga Enough to Keep You Fit?" *Yoga Journal*, September–October 2002.

Bear, Mark F., Barry W. Connors, and Michael A. Paradiso. *Neuroscience: Exploring the Brain*, 3rd ed. Philadelphia: Lippincott Williams & Wilkins, 2006.

Behanan, Kovoor T. *Yoga: A Scientific Evaluation*. New York: Macmillan, 1937.

Bhole, M.V., P.V. Karambelkar, and S.L. Vinekar. "Underground Burial or Bhugarbha Samadhi (Part II)." *Yoga Mimansa*, vol. 10, no. 2, October 1967, pp. 1–16.

Braid, James. *Observations on Trance: or, Human Hibernation*. London: John Churchill, 1850.

Briggs, George Weston. *Gorakhnath and the Kanphata Yogis*, reprint of the 1938 ed. Delhi: Motilal Banarsidass, 1989.

Budilovsky, Joan, and Eve Adamson. *The Complete Idiot's Guide to Yoga*, 3rd ed. Indianapolis: Alpha Books, 2003.

Burton, Hunphrey. *Yehudi Menuhin: A Life.* Boston: Northeastern University Press, 2001.

Chaya, M.S., A.V. Kurpad, H.R. Nagendra, et al. "The Effect of Long-term Combined Yoga Practice on the Basal Metabolic Rate of Healthy Adults." *BMC Complementary and Alternative Medicine*, vol. 6, no. 28, published online August 31, 2006, www.biomedcentral.com/1472-6882/6/28.

Choudhury, Bikram. *Bikram Yoga: The Guru Behind Hot Yoga Shows the Way to Radiant Health and Personal Fulfillment.* New York: HarperCollins, 2007.

Christiansen, K. "Behavioral Correlates of Testosterone," in Eberhard Nieschlag and Hermann M. Behre, eds., *Testosterone: Action, Deficiency, Substitution*, 3rd ed. Cambridge, England: Cambridge University Press, 2004.

Cooper, Kenneth H. *Aerobics.* New York: Bantam, 1968.

Coulter, H. David. *Anatomy of Hatha Yoga: A Manual for Students, Teachers, and Practitioners.* Honesdale, PA: Body and Breath, 2001.

Degler, Teri. *Fiery Muse: Creativity and the Spiritual Quest.* Toronto: Random House, 1996.

De Michelis, Elizabeth. *A History of Modern Yoga: Patanjali and Western Esotericism.* London: Continuum, 2005.

Eliade, Mircea. *Yoga: Immortality and Freedom.* Princeton, NJ: Princeton University Press, 1990.

Evans-Wentz, W.Y., ed. *The Tibetan Book of the Dead.* New York: Oxford University Press, 2000.

Feuerstein, Georg, and Larry Payne. *Yoga for Dummies.* Indianapolis: Wiley, 1999.

Fishman, Loren, and Carol Ardman. *Relief Is in the Stretch: End Back Pain Through Yoga.* New York: Norton, 2005.

———, and Eric L. Small. *Yoga and Multiple Sclerosis: A Journey to Health and Healing.* New York: Demos Medical Publishing, 2007.

———, and Ellen Saltonstall. *Yoga for Arthritis: The Complete Guide.* New York: Norton, 2008.

Garbe, Richard. "On the Voluntary Trance of Indian Fakirs." *The Monist*, vol. 10, no. 4, July 1900, pp. 481–500.

Gharote, Mandhar L., and Manmath M. Gharote. *Swami Kuvalayananda: A Pioneer of Scientific Yoga and Indian Physical Education.* Lonavla, India: The Lonavla Yoga Institute, 1999.

Green, Elmer and Alyce Green. *Beyond Biofeedback.* New York: Delacorte Press, 1977.

Grof, Stanislav, and Christina Grof, eds. *Spiritual Emergency: When Personal Transformation Becomes a Crisis.* New York: Putnam, 1989.

Gune, Jagannath G. *Popular Yoga Asanas,* reprint of the 1931 book. Rutland, VT: Charles E. Tuttle, 1974. The book gives the author as Swami Kuvalayananda.

Isaacs, Nora. "The Yoga Therapist Will See You Now." *New York Times,* May 10, 2007, p. 10.

Iyengar, B.K.S. *Light on Yoga.* New York: Schocken, 1979.

———. *Iyengar: His Life and Work.* Porthill, ID: Timeless Books, 1987.

———. *Astadala Yogamala,* vol. 1. New Delhi: Allied Publishers, 2006.

———. *Light on Pranayama: The Yogic Art of Breathing.* New York: Crossroad Publishing, 2006.

Jacobson, Edmund. *Progressive Relaxation: A Physiological and Clinical Investigation of Muscular States and Their Significance in Psychology and Medical Practice.* Chicago: University of Chicago Press, 1931.

Jung, C. G. "Psychological Commentary," in W. Y. Evans-Wentz, ed., *The Tibetan Book of the Dead.* New York: Oxford University Press, 2000.

———, Aniela Jaffé, ed., Richard and Clara Winston, trans., *Memories, Dreams, Reflections.* New York: Vintage, 1989.

Khalsa, Shakta Kaur. *Kundalini Yoga: Unlock Your Inner Potential Through Life-Changing Exercise.* New York: Dorling Kindersley, 2001.

Kolata, Gina. *Ultimate Fitness: The Quest for Truth About Exercise and Health.* New York: Picador, 2004.

Komisaruk, Barry R., Carlos Beyer-Flores, and Beverly Whipple. *The Science of Orgasm.* Baltimore: Johns Hopkins University Press, 2006.

Krishna, Gopi. *The Biological Basis of Religion and Genius.* New York: Harper & Row, 1972.

———. *Kundalini: The Evolutionary Energy in Man.* Boston: Shambhala, 1997.

Kuvalayananda, Swami. *Popular Yoga Asanas,* reprint of the 1931 book. Rutland, VT: Charles E. Tuttle, 1974.

Lasater, Judith Hanson. *Yogabody: Anatomy, Kinesiology, and Asana.* Berkeley: Rodmell Press, 2009.

Lum, L.C. "Hyperventilation Syndromes," in Beverly H. Timmons and Ronald Ley, eds., *Behavioral and Psychological Approaches to Breathing Disorders.* New York: Kluwer, 1994, pp. 113–123.

Mallinson, James, trans. *Gheranda Samhita*. Woodstock, NY: YogaVidya .com, 2004.

Masters, William H., and Virginia E. Johnson. *Human Sexual Response*. Boston: Little, Brown, 1966.

McArdle, William D., Frank I. Katch, and Victor L. Katch, *Exercise Physiology: Nutrition, Energy, and Human Performance*, 6th ed. Philadelphia: Lippincott Williams & Wilkins, 2007.

McCall, Timothy. *Yoga as Medicine: The Yogic Prescription for Health and Healing*. New York: Bantam, 2007.

Ogden, Gina. *Women Who Love Sex: Ordinary Women Describe Their Paths to Pleasure, Intimacy, and Ecstasy*. Boston: Trumpeter, 2007.

Osborne, William G. *The Court and Camp of Runjeet* [sic] *Sing* [sic], reprint of 1840 ed. Karachi: Oxford University Press, 1973.

Paul, N. C. *A Treatise on the Yoga Philosophy*. Benares: Recorder Press, 1851.

Payne, Larry, and Richard Usatine. *Yoga Rx: A Step-by-Step Program to Promote Health, Wellness, and Healing for Common Ailments*. New York: Broadway Books, 2002.

Robin, Mel. *A Physiological Handbook for Teachers of Yogasana*. Tucson: Fenestra Books, 2002.

———. *A Handbook for Yogasana Teachers: The Incorporation of Neuroscience, Physiology, and Anatomy into the Practice*. Tucson: Wheatmark, 2009.

Porth, Carol Mattson. *Essentials of Pathophysiology*. Philadelphia: Lippincott Williams & Wilkins, 2007.

Ruiz, Fernando Pagés. "Krishnamacharya's Legacy." *Yoga Journal*, May–June 2001, pp. 96–101, 161–168.

Samuel, Geoffrey. *The Origins of Yoga and Tantra: Indic Religions to the Thirteenth Century*. New York: Cambridge University Press, 2008.

Sannella, Lee. *The Kundalini Experience: Psychosis or Transcendence?* Lower Lake, CA: Integral Publishing, 1992.

Schievink, Wouter I. "Spontaneous Dissection of the Carotid and Vertebral Arteries." *New England Journal of Medicine*, vol. 344, no. 12, March 22, 2001, pp. 898–906.

Schmidt, T., A. Wijga, A. Von Zur Mühlen, et al. "Changes in Cardiovascular Risk Factors and Hormones During a Comprehensive Residential

Three-Month Kriya Yoga Training and Vegetarian Nutrition." *Acta Physiologica Scandinavica, Supplement*, vol. 640, 1997, pp. 158–62.

Schwartzstein, Richard M., and Michael J. Parker. *Respiratory Physiology: A Clinical Approach*. Philadelphia: Lippincott Williams & Wilkins, 2006.

Shamdasani, Sonu, ed. *The Psychology of Kundalini Yoga: Notes of the Seminar Given in 1932 by C. G. Jung*. Princeton, NJ: Princeton University Press, 1999.

Shaw, Beth. *Beth Shaw's YogaFit*. Champaign, IL: Human Kinetics, 2000.

Siegel, Lee. *Net of Magic: Wonders and Deceptions in India*. Chicago: University of Chicago Press, 1991.

Singleton, Mark. *Yoga Body: The Origins of Modern Posture Practice*. New York: Oxford University Press, 2010.

Sjoman, Norman E. *The Yoga Tradition of the Mysore Palace*. New Delhi: Abhinav Publications, 1999.

Taimni, I. K. *The Science of Yoga*. Wheaton, IL: Quest, 1972.

Taylor, Jill Bolte. *My Stroke of Insight: A Brain Scientist's Personal Journey*. New York: Viking, 2006.

Udupa, K. N. *Stress and Its Management by Yoga*. New Delhi: Motilal Banarsidass, 2000.

Urban, Hugh B. *Tantra: Sex, Secrecy, Politics, and Power in the Study of Religion*. Berkeley: University of California Press, 2003.

Vishnudevananda, Swami. *The Complete Illustrated Book of Yoga*. New York: Bell Publishing, 1960.

White, David Gordon. *Kiss of the Yogini: "Tantric Sex" in Its South Asian Contexts*. Chicago: University of Chicago Press, 2006.

———. *Sinister Yogis*. Chicago: University of Chicago Press, 2009.

———, ed. *Tantra in Practice*. Princeton, NJ: Princeton University Press, 2000.

World Health Organization. "Depression." www.who.int/mental_health /management/depression/definition/en.

Yogani. *Advanced Yoga Practices: Easy Lessons for Ecstatic Living*. Nashville: AYP Publishing, 2004.

Acknowledgments

My thanks go first and foremost to the scientists and other specialists who made this book possible. Their courtesy and patience—in some cases over a number of years—helped illuminate a subject so murky and complicated that I despaired at times of figuring it out. Although I now question much about the culture of modern yoga, I hope that anyone who feels uneasy with my skepticism will nonetheless see my reporting as thorough and fair. As always with authorship, I alone am responsible for any errors or significant omissions.

Initially, I saw this book as a nine-month wonder in which I would pick the low-hanging fruit and go merrily on my way. Five years later, I have accumulated a large number of debts.

For early advice and encouragement, I offer heartfelt thanks to Joseph S. Alter, Charlotte Bacon and Brad Choyt, R. Barker Bausell, Carolyn Marks Blackwood and Greg Quinn, Ingrid and Walter Blanco, William C. Bushell, John Eastman, Jack England, Owen Gingerich, Ann Godoff, Daniel Goleman, John Horgan, Alan Lightman, Gary Rosen, and Patricia L. Rosenfield.

In Kolkata, I am much indebted to Ashim Mukerji of the National Library, to P. Thankappan Nair, the journalist mentioned in chapter 1, to Binoy Roy, the former librarian of the University of Calcutta, and to the obliging staffs of the Asiatic Society and the Bengal Academy of Literature.

In Mumbai, I battled monsoon flooding and dead taxicabs to laugh my head off with Madan Kataria and his good friends.

In Lonavla, special thanks go to Subodh Tiwari and Swati Deshpande of the Kaivalyadhama Yoga Institute (Gune's ashram) and to Manmath M. Gharote of the Lonavla Yoga Institute and his colleagues. In Bangalore, H. R. Nagendra came to my rescue. In Washington at the Indian embassy, Nikhilesh M. Dhirar and his colleagues worked hard to track down an unforthcoming fact.

Graciously over the years, Priscilla Walker lent her files and knowledge to unraveling the tale of Yogananda and Basu Kumar Bagchi. So, too,

Katharine Webster helped with Swami Rama. Randi Hutter Epstein cast light on yoga history, and Carolyn Griffin on a sudden death.

For assistance in learning something of contemporary Tantra and Kundalini, many thanks to Bob Boyd, Michael Bradford, Joan Bridges, Jennifer Clark, Jana Dixon, Judy Harper, David Lukoff, Stuart Sovatsky, and Lisa Paul Streitfeld. Much gratitude as well to Ilse Mohn for a commercial insight and to Mary Roach for general intelligence on the relationship between sex and modern yoga.

Gene Kieffer lent his considerable resources to helping me better understand Kundalini in general and Gopi Krishna in particular.

Walter Blanco and James Anderson helped inspire the Muse chapter with their lively discussion of Sonny Rollins.

For acts of kindness and assistance, thanks to Angela Babb of the American Academy of Neurology, to Lynn Butler, Laura Tatum, Nancy Lyon, and Angelyn Singer of Yale University, to Dennis Campbell and Patricia Gallagher of the New York Academy of Medicine, to Linda Cuthbertson and Pamela Forde of the Royal College of Physicians, to Daniel DeBehnke and Terry Modrak of the Medical College of Wisconsin, to Janet Faubert and Myrna Filman of the Institute for Consciousness Research, to Daisy Franco of the American Medical Association, to Emil Frantík of the National Institute of Public Health in Prague, to Sharon Gardner of the University of Michigan, to Diane Gray-Reed of the Pacifica Graduate Institute Research Library, to Mary Guillemette of the *Archives of Physical Medicine and Rehabilitation*, to Stephanie Hawthorn of the British Medical Association, to Derek Johnson of the Berkshire Medical Center, to Cindy Kuzma of the *Journal of the American Medical Association*, to Vicky Leonard and Scott Wolfson of the Consumer Product Safety Commission, to Robert Love of the Columbia University Graduate School of Journalism, to John McKenzie of Sumner McKenzie, to Renate Myles of the National Institutes of Health, to Natalya Podgorny of the Himalayan Institute, to Melanie Walker of the University of Washington, and to Susan Weill of *Time* magazine.

Calm at the center of the storm? Yes, thanks to Ellen Patrick of Yoga Sanctuary in Mamaroneck, New York, to Athina Pride of the Infinite Yoga Center in Larchmont, New York, and to Jessica Thompson of the Yoga Loft in Bethlehem, Pennsylvania. Franklin Shire was as much fresh breeze as yoga instructor.

Good friends practiced the discipline of hearty encouragement and good cheer. Thanks to Jane Elkoff and Peter Gregersen, Rima Grad and Neil Selinger, Abby Gruen and Bob Graubard, Marnie Inskip and Dan O'Neill, Sophie and Tom Kent, Martha Upton and Peter Davis, Catherine and Stuart Wachs, and Sarit and Harry Wall.

Friends at the Larchmont Public Library—icons of professional courtesy and restraint—put up with years of hectoring. Many thanks to Frank Connelly, Paul Doherty, Nancy Donovan, Liam Hegarty, and June Hesler, as well as the many unseen hands of the Westchester Library System.

Colleagues at *The New York Times* offered advice, support, and considerable help in gathering papers and other materials. Thanks to Lawrence K. Altman, Pam Belluck, Toby Bilanow, Benedict Carey, Laura Chang, David Corcoran, Henry Fountain, Denise Grady, Erica Goode, James Gorman, Leslie Kaufman, Soo-Jeong Kang, Gina Kolata, Mireya Navarro, Tara Parker-Pope, David Sanger, Elaine Sciolino, Barbara Strauch, and Nicholas Wade. Thanks, too, Gina, for *Ultimate Fitness*.

A number of specialists, colleagues, and family members took time to comment on all or part of the manuscript and helped improve it in countless ways. Many thanks to Chris Arrington, Brenda Berger, Carole Anne Broad, Charles A. Broad Jr., Mary Broad, Christina Bryza, Bobby and Lindsey Clennell, Jane Elkoff, Daniel Goleman, Randi Hutter Epstein, Jane Keogh Kelly, Sharon Maier, Jarl Mohn, Luis Parada, Franklin Shire, Mark Singleton, Stuart Wachs, and Nicholas Wade. Special thanks to Jarl and Nicholas for thoughtful suggestions.

My illustrator, Bobby Clennell, teaches yoga in New York City, has studied with Iyengar in India, and has authored two yoga books of her own—*The Woman's Yoga Book* and *Watch Me Do Yoga*, for children. She and her model, Lisa Rotell, performed the magic of turning abstractions into revealing images both instructive and elegant.

My agent, Peter Matson, offered encouragement, good advice, and unfailing good humor throughout the book's ups and downs. I find myself ever more in your debt, Peter. Thank you.

My gratefulness to Alice Mayhew, my editor at Simon & Schuster, goes beyond words. Her thoughtful advice and relentless enthusiasm brought this book into the world. We've done other projects together but this one required unusual skill and sensitivity. Thank you, Alice— and thanks to your many gifted associates at Simon & Schuster, most especially Roger Labrie. Thanks also to publisher Jonathan Karp, Irene Kheradi, Nancy Inglis, Renata Di Biase, Julia Prosser, and Rachel Bergmann. No turn of phrase can express the depth of my gratitude.

Over the decades, I have learned to rely on Tanya Mohn, my wife, and my three children, Max, Isabelle, and Juliana, for endless support, forbearance, and love in this crazy process of book writing. Thanks, guys. You are my prana. And thank you especially, Tanya. You are not only my life energy but my ethicist and guru—among other roles. Your counsel and wisdom typically become the best parts of me. Namaste.

Finally, I'd like to honor the memory of Nancy, a much-loved sister lost to cancer. More than forty years ago she played an important role in getting me interested in yoga and pursuing it as a life discipline. For that, Nancy, you will always hold a special place in my heart.

William J. Broad
Larchmont, New York
December 2011

Note: Page numbers in *italics* refer to illustrations.

About the Author

WILLIAM J. BROAD has practiced yoga since 1970. A senior writer at *The New York Times*, he has written hundreds of front-page articles and won every major award in print and television during more than thirty years as a science journalist. With *Times* colleagues, he has twice won the Pulitzer Prize, as well as an Emmy and a duPont. He is the author or coauthor of seven books, including *Germs: Biological Weapons and America's Secret War* (Simon & Schuster, 2001), a number one *New York Times* bestseller. He has three adult children and lives with his wife in the metropolitan New York area, where he enjoys doing Sun Salutations.